Daily Ethics

Creating Intentional Practice for Behavior Analysts

Daily Ethics

Creating Intentional Practice for Behavior Analysts

Tyra P. Sellers

Emily A. Patrizi

Sarah Lichtenberger

KeyPress Publishing
www.keypresspublishing.com

Authors: Tyra P. Sellers, Emily A. Patrizi, Sarah Lichtenberger

Daily Ethics: Creating Intentional Practice for Behavior Analysts

Published by: KeyPress Publishing
Publisher: Alice Darnell Lattal
Brand Integrity: Jana Burtner
Production Manager: Adele Hall
Editors: Ashley Johnson and Stefanie Carr
Designer: Kim Harding

ISBN 979-8-9886548-5-8

Library of Congress Control Number: 2024942357

Published in Melbourne, Florida

Distributed by:
ABA Technologies, Inc.
930 South Harbor City Blvd, Suite 402
Melbourne, FL 32901
www.abatechnologies.com

Also from Tyra Sellers and KeyPress Publishing:

Sellers and LeBlanc (2022):
The New Supervisor's Workbook:
Success in the First Year of Supervision

LeBlanc and Sellers (2022):
The Consulting Supervisor's Workbook:
Supporting New Supervisors

This book is dedicated to YOU. Yeah, that's right. We dedicate this book to the person reading this who must navigate through the ethical gray shades on a daily basis. Thank you.

Tyra Sellers

I dedicate this book to my partner, Adam; my son, Maxwell; and my daughter, Charlotte, who have loved me even in my most unlovable moments. And to two of my besties, Dr. Linda LeBlanc and Dr. Katie Harris, who provide me with just the right combination of intellectual conversation and pants-peeing shenanigans. Finally, to Cleo Tarbox, who taught me the term "scarycool," which helped me lean into many an uncomfortable ethics topic and dilemma.

Emily Patrizi

I dedicate this book to my two incredible kids, who make me better in every way. To my daughter, Elianna, who moves through life with a tenacity and grace that take my breath away. And to my son, Matthew, who embraces life with exhilarating joy that reminds me to live each day with gratitude. To my parents, my greatest supporters, and my grandparents, who taught me to experience the beauty in everything.

Sarah Lichtenberger

I dedicate this book to Lee, who joins me in every new adventure with enthusiasm and unwavering support. To my family, Mark, Raziya, and Emmett, who have always encouraged me. And to Amber and Abby, for answering every phone call, practicing every difficult conversation, and always cheering me on.

Contents

Preface

A Bit About How *Daily Ethics: Creating Intentional Practice for Behavior Analysts* Came to Be: This book has been some time in the making. I (Tyra) dreamed up ideas for something more than a textbook while teaching ethics at Utah State University (USU). My course was informed by my many years in clinical practice and my time in law school. The available textbooks did a fine job of introducing the profession's history and reviewing the available professional code of ethics. But I worried that a single course was insufficient to prepare my students for what was to come. I longed for something practical that professionals could use after an ethics course to support them in interacting with ethics regularly as they experienced real-work ethical quandaries and dilemmas throughout their careers. My time as Director of Ethics at the Behavior Analyst Certification Board® (BACB®) only solidified my desire for such a resource. I wanted something that made it clear that applied professional ethics is not black and white and, instead, requires active analysis of uncomfortable content that considers multiple perspectives—a sort of field guide that practitioners could use to enhance their skills and return to when needed.

I talked about the idea with Emily and Sarah over the years. They seemed to think the idea was a good one. Throughout our conversations, they both added important perspectives and new considerations that enhanced the focus of the book. The three of us worked collaboratively to develop what we hope is a useful tool that professional behavior analysts can use to develop a deeper relationship with ethics content and to create a daily practice of looking for and thinking about ethics. Not in an anxious way, but in a natural way that informs how they move through their daily professional activities.

A Bit About Tyra Sellers: What to say? I am forever grateful that each of my attempts to veer from the path of a career in applied behavior analysis (ABA) failed. I am lucky that my education is a smorgasbord of psychology, anthropology, philosophy, law, special education, and ABA. I am luckier that my journey has been winding and varied, taking me from special education to clinical and consulting services, then sliding into leadership positions, then looping around to time teaching and conducting research as university faculty at USU, then veering into supporting the ethics department at the BACB, then returning to consultation, and finally, at the time of writing, changing lanes slightly to serve at the Association of Professional Behavior Analysts (APBA). And I am luckiest to have been shaped by the loveliest clients, caregivers, students, supervisees, trainees, colleagues, supervisors, mentors, and the dearest of friends along the way. The offerings in this book come from my heart and the myriad mistakes I've made throughout my 30+ year career—I hope they prove helpful for even one reader.

A Bit About Emily Patrizi: The words in this book reflect our collaborative experiences, and I am forever grateful for both the experiences and the opportunity to share these lessons on paper. My early beginnings were in the world of education and direct care in group homes. Throughout the years, I had the opportunity to support across a variety of clinical and operational roles within the ABA field, with increasing scope and responsibility; establish an ABA program to expand an interdisciplinary model; and serve as a coach with the BACB (one of my fondest roles). These experiences taught me countless lessons, and who I am today is the result of the colleagues, leaders, friends, clients, and families who walked with me along the way. Of all these incredible teachers, my greatest teacher has been cancer. It is dually the best and worst experience that I have lived through and has given me a renewed lens of humility and gratitude that has changed who I am and how I live my life. I have realized that nothing is ever as bad or as good as it seems, and that laughter is almost always the answer. I believe that we are here to be kind and do what is right, and that is how I strive to live every day, even though I often fail miserably. When we show up with the intention to learn and be better, we get better in grace and in time. Thank you for being a part of this journey with us.

A Bit About Sarah Lichtenberger: Who knew that an eventful summer working in a group home and one simple web search could have led me on such an adventure? I'm grateful that my experiences as a direct support professional, student, supervisee, practitioner, and supervisor have led me here and for the many people I worked with along the way who participated in my journey (my coauthors included). My experiences in clinical work and working in the ethics department at the BACB gave me a new perspective on how we can support each other in this profession and create a community built on education and growth. For me, this book comes from wanting to share the opportunities for growth and forgiveness that have been shared with me from others along the way (and a little from myself). Know that this book comes from a place of accepting that we cannot be perfect and that we will always keep learning, with the hope that I can share a small part of what I've learned and experienced with others to support you on your journey. And an additional plus—I finally got to use my English degree!

Acknowledgments

We are grateful to our publisher, KeyPress, for the opportunity to make this book a reality. To our editor, Ashley Johnson, for her kind *and* honest feedback and always thoughtful ideas; we are forever indebted. Finally, a heartfelt thank you to each and every author of a textbook, scholarly article, or research article focused on ethics in ABA—we could not have written this without having been informed by their work.

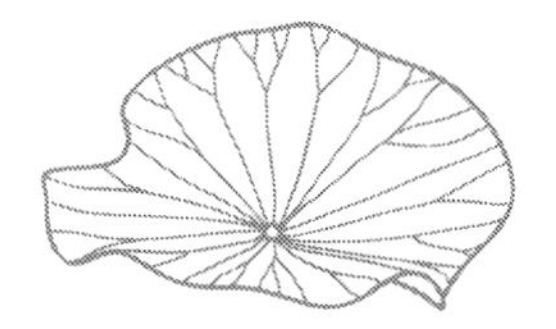

Chapter 1

Introduction—Get the Context

Welcome! We are so excited that you are interested in developing your skills around professional ethics in the profession of applied behavior analysis (ABA) services. Did you know that we are part of a very young profession? For example, our primary certifying body, the Behavior Analyst Certification Board (BACB), was founded in 1998 and began certifying folx in 1999 (Sellers et al., 2023). And my goodness, how we have grown! At the time of writing this book, there are a reported 64,626 Board Certified Behavior Analysts® (BCBAs®) and 5,431 Board Certified Assistant Behavior Analysts® (BCaBAs®). For some context, it took 13 years to reach 10,000 BCBAs and another 6 years to reach 32,000 BCBAs (BACB, n.d.).

Since you're reading this book, you're clearly aware that behavior analysts have ethics requirements that we must be aware of and comply with. These requirements might come from your organization, licensure board, or certification body. It's also likely that you're certified or credentialed (or hoping to become certified) by the BACB and are aware of its specific ethics requirements. Oh, don't get scared—this is not a textbook or a history book. But it can be helpful to review the foundation and current landscape of our profession related to ethics.

The BACB's first version of an ethics code—*Guidelines for Responsible Conduct for Behavior Analysts*—was adopted in 2001 (BACB, 2001; Sellers et al., 2023). Updates have been made over time, bringing us to the most recent versions at the time of writing: *Ethics Code for Behavior Analysts* (BACB, 2020a) and RBT® *Ethics Code (2.0)* (Registered Behavior Technician®; BACB, 2021d). These documents describe the scope of the code and list the principles and standards that ensure we are protecting consumers and the profession of behavior analysis by engaging in ethical practice.

Even though we all have to take an ethics course in graduate school and complete required continuing education focused on ethics, we need to continually build our skills around ethics. Why? Well, that course and those ethics talks and workshops are likely not enough to ensure that we are keeping ethics at the forefront of our practices. Not convinced? Okay, well maybe some data will help.

There seem to be some trends in the profession related to the most frequent areas of actionable ethics violations. In a white paper published in 2018 (BACB, 2018b), the three categories with the highest number of actionable violations are "improper or inadequate supervision or delegation," "failure to report or respond to the BACB as required," and "professionalism and integrity" (p. 5). A more recent white paper published in 2023 (BACB, 2023b) shows some similarities. The most frequent areas of potential and actionable violations listed are in the *same three areas* as in the 2018 paper, as well as areas including "improper discontinuation or transition of services/service interruptions," "nonsexual multiple or exploitive relationship(s)," and "failure to maintain adequate or accurate records" (p. 5).

These resources and information can help us identify what areas we need to focus our attention on when evaluating our own practice and when training and supervising others. They also highlight that ethics violations are occurring, despite our graduate coursework and required continuing education efforts, and that they are occurring related to some very common behaviors, like supervision, multiple relationships, and professionalism. That's where this book comes in. This book bridges the gap between what you learned in a course however many years ago and the sporadic professional development ethics content you contact. This book helps you place ethics into the daily context of your professional life.

Okay, so now that we've covered a bit about the history of our professional ethics and a bit about the current landscape of ethics in our profession, let's talk about applied ethics and set the context for this book. It can be easy to think about ethics in terms of big dilemmas, instances where the stakes are high, where people could get really hurt, and where laws could be broken. We can also be tricked into thinking that a course in ethics and an understanding of our profession's ethical obligations are enough to prepare us to handle ethical dilemmas that we will undoubtedly face. And for many of us, our instruction focused on learning how to identify ethics violations presented in scenarios.

No doubt, ethics is involved in examples of harming clients, engaging in purposeful billing fraud, or exploiting your team. But ethics is also about, in fact *most frequently* about, our day-to-day activities. It's about the "ordinary questions of daily life" (Longstaff, 2017, p. 27). For practicing behavior analysts, ethics shows up in those big dilemmas, but also in those everyday questions and choices about things like accepting a client or not, choosing a data collection system, deciding to accept a request to connect on social media from a trainee or caregiver, explaining an assessment procedure to a caregiver, selecting an intervention, or providing corrective feedback.

It's also the case that our ethics coursework and familiarity with our profession's codes of ethics provide foundational knowledge about relevant ethics requirements and challenges we're likely to face. But is that enough? Pope and Vasquez (2016) caution that we cannot rely on our professional ethics codes to answer questions, tell us how we should feel, or even tell us how to respond in a given context—that is up to us. Our professional codes of ethics are necessary to provide protection to consumers, practitioners, and the profession and to guide us in "right" and "wrong" practices (Pope & Vasquez, 2016). But as practitioners, we must develop the repertoires that will support us in actively struggling with the unique, complicated, and sometimes competing situations that we will face in the context of serving others.

Finally, reading scenarios that present ethics violations can be an effective way to learn to identify related ethics standards from codes of ethics, thereby increasing familiarity with our ethics obligations. Those scenarios can also be used to support us in outlining what could have been done to improve the situation. However, it's likely that at least some, if not most, of the situations involving ethical dimensions that creep up in our practice are difficult to detect in the moment. These situations might be tricky to detect before we contact any of the problematic consequences or before any harm occurs that might draw attention to the issue. This might be especially true if the bulk of our training has focused on discriminating violations that have occurred, rather than identifying red flags and risk factors. We all ought to spend some time learning to identify the contexts and behaviors that might increase our "ethical blindness," or temporary inability to discriminate and describe the ethical dimensions of our behavior, choices, and practices (Palazzo et al., 2012).

What Are You Getting Yourself Into by Reading This Book?

Our work as behavior analysts is complex, ever-changing, and threaded through and through with ethics choices, pitfalls, and dilemmas. This book is a direct (and supportive) call to actively engage with ethics content daily. We're asking you to pause, look around, and bring into focus the fact that ethics is embedded in all activities of a practicing behavior analyst. We believe that ethics is not something to pull off a shelf and dust off when faced with an ethical dilemma, or worse, after you discover that an ethics violation has happened. On the contrary, ethics is something that should be at the forefront of our daily activities. Think of ethics as a muscle that you develop and strengthen with intentional practice. Why? Well, because this stuff can be difficult, and even hard to see. And because choices and situations that involve ethical dilemmas can be uncomfortable and scary. They can evoke go-to responses like avoidance and escape, which almost always make the situation worse.

Mary Gentile (2012) talks about developing moral or ethical muscle memory. Having a strong ethical muscle memory, like an accomplished athlete, increases our chances of proactively behaving in ways aligned with our values and ethics requirements when we're faced with ethical choices. According to Gentile, we've got to engage in applied training by actively strategizing, preparing, and practicing how we'll respond in these situations. In behavior-analytic terms, we have to strengthen desired and effective responses to ethics choices and dilemmas in the practice context so that when the stimuli present themselves in the natural context, we're likely to engage in the desired responses, as opposed to responding impulsively or avoiding the situation.

Okay, we mentioned engaging in *intentional practice*, and by that, we mean committing to engaging in daily ethics practice. We purposefully chose the term *practice*—a verb that means "to perform or work at repeatedly so as to become proficient" (Merriam-Webster, n.d.-b). Getting better at something requires practice. For us, thinking about ethics as a practice embodies that we'll get it wrong sometimes, that we'll learn from our mistakes, AND that we'll keep going in our practice activities! To that end, this book presents ethics topics in biweekly pairs with prompts and activities to support you on your journey to develop a daily practice of attending to ethics.

In this daily practice, we're going to ask you to focus primarily on yourself, on deepening your understanding of and increasing your comfort level with applied ethics. Identifying when someone else has done something we think is wrong or that we don't like is easy. But shining a light on our own thoughts, feelings, biases, and behavioral patterns is a bit trickier and a bit *ouchier*. Real talk—if we want to increase the chances that we'll behave ethically every day, and that we'll notice risks in our behavior and environment, then we should focus on the only person's behavior that we can actually control—our own. So, we're asking that you lean into the discomfort you might experience and focus on yourself. We're also asking that you acknowledge that every single behavior analyst (yep, you and all three of us included) is at risk of finding themselves in an ethical pickle. There's no amount of "being a good person" or "loving ethics" that will function as protective armor against contacting ethics issues, because they're present in everything we do.

Of course, throughout the book, we'll also include strategies for identifying and addressing concerns with others. But, by putting in the work to build your ethical muscle memory, we hope that your day-to-day practices might be better aligned with our professional ethics, that you might be better prepared to spot potential ethics risks and actively avoid and manage them, and that you'll have functional and effective responses at peak form to assist you in navigating the ethical dilemmas and violations that you might encounter throughout your career.

It's also worth noting ahead of time what types of examples you're going to find in the biweekly pairs. It's a beautiful thing that behavior analysts implement ABA-based services across a wide range of subspecialty practice areas (e.g., behavioral gerontology, behavioral sports psychology, business, brain injury, health and fitness, organizational behavior management, public health, substance use disorders). However, at the time of writing, most practicing behavior analysts work in the context of providing ABA-based therapies to individuals with intellectual and developmental disabilities, most notably with autistic individuals, in homes, clinics, and schools. So, to make the content as relatable as possible to practitioners, we decided to create scenarios that reflect that main practice area. The content is still relevant to those practicing in different areas; you just might have to translate some of the content in the examples to your own practice focus or setting.

What Are You *Not* Getting Yourself Into by Reading This Book?

First and foremost, you're not getting a textbook. Generally speaking, a textbook provides comprehensive information about a topic for the purpose of teaching about that topic. This book is neither comprehensive, nor is it focused on teaching you about the topic of ethics. As a profession, we're lucky to have many chapters in behavior analysis textbooks, and full textbooks (Bailey & Burch, 2019; Bailey & Burch, 2020; Bailey & Burch, 2022; Beirne & Sadavoy, 2021; Brodhead et al., 2022; Flowers, 2023; Kelly, Shraga, & Bollinger, 2023; LeBlanc & Karsten, 2024; Sush & Najdowski, 2022) focused on teaching about ethics. You're also not getting yourself into a description of each of the standards in our profession's ethics codes or a compendium of practice scenarios (although we do present scenarios as a prompt to apply the content), because we already have great books that do both of those things.

Instead, think of this book more as a field manual that provides practical descriptions and prompts for how to engage with and apply the ethics knowledge you already have in a meaningful and active manner. And like a field manual, we hope it becomes a resource that you can come back to again and again to support your daily practice, give depth to your supervisory practice, and help when you find yourself faced with an ethics challenge.

Ready to Go on a Journey With Us?

Because ethics can be scary and overwhelming—both things that make most of us want to run in the other direction—we offer this book as an invitation to lean into these topics directly, but slowly, over the course of a year. We offer this book as a starting place for a lifelong journey of active engagement with ethics topics. This is an invitation to commit to the work of developing ethical muscle memory with focused, daily practice with a variety of ethics topics and exercises. Some of the work will be easy and obvious. Some of the work will be challenging and uncomfortable. So, thanks for accepting our invitation and taking this journey with us. We'll be right there with you, cheering you on and reminding you that you're doing a great job!

Chapter 2

How to Use This Book—Get Your Bearings

As with any journey, before you get started, it's good to have a general lay of the land. You've got the context for this journey—to commit to daily engagement with ethics topics. Let's review the basic road map to help us navigate it. The book is divided into five main sections:

1. Introduction—Get the Context
2. How to Use This Book—Get Your Bearings
3. But First—Get Yourself Prepped
4. Week by Week—Get Going!
5. Appendices

Here's the big picture: After this chapter, you'll read Chapter 3: "But First—Get Yourself Prepped," which will help bring your attention to some of the overarching themes and some skills you're going to use and continue to develop for this journey. Themes and related skills include compassion, self-reflection and self-care, cultural humility and responsiveness, values, and having difficult conversations. After that, you'll get started with the weekly ethics topics. Before we review what you'll find in the biweekly pair sections, let's review the week-by-week schedule and how you might approach using this book.

Week-by-Week Schedule

We have mapped out the book across 52 weeks, but that doesn't mean that you need to start this journey on January 1st. On the contrary, we invite you to jump into this journey with us wherever you are right now. We'll all start with Weeks 1 & 2 together. If this is your first time reading the book, we recommend that you start with the first biweekly pair and go through the book in order to help you stay committed to engaging in daily practice. If you decide not to do that, we strongly encourage you to read Chapter 3 and the first biweekly pair so that you have a strong foundation. Weeks 1 & 2 prepare you for the daily practice that you'll encounter in the remaining biweekly pairs.

In Week 1, we'll introduce you to the practice of thinking about and reflecting on ethics and noticing elements related to ethics in your professional behavior and in the workplace. These practices need to come from a place of non-judgment, compassion, and collaborative learning. In Week 2, we'll prepare you for the practices of engaging in meaningful, and sometimes challenging, conversations with others and taking action to apply the content to your daily practice.

Whether you go through the content in order, as we suggest, or you plan on jumping around, it's a good idea to flip through the biweekly pairs, so you get a feel for the topics covered. We also recommend that once you get to the first biweekly pair at Weeks 3 & 4, you develop the habit of scanning the biweekly pair content the day before you get started (i.e., Sunday), or the Monday of the first week. This is important because some of the content can be more challenging than others. And some topics can be enhanced or complicated by what's going on in your personal or work life. Scanning the biweekly pair ahead of time gives you the chance to decide if you're ready to roll up your sleeves and forge ahead on your journey or need to make the decision to pull over and take a break until you're ready and able to reengage.

If you find that you need to pause, that's totally fine; be kind to yourself. In fact, we build in some breaks for you! After three biweekly pairs, you'll find a week titled "Time to Rest & Take Care of Yourself." Think of these weeks as rest stops along your path to daily ethics practice. Following the rest week, you'll find a week titled "Time to Reconnect With Your Daily Practice" that prompts you to ease back into your daily practice of focusing on ethics. But this is your journey, and you control the pace. If you do pause things, just make sure to get back on the road as soon as you are able. We'll be here, patiently waiting for your return.

You might be reading through on your own, which is a perfectly appropriate way to use this book. In fact, it's a great approach to take the first time you interact with the content. You could also read through the book with a friend, with a mentor or supervisor, or in a book club with several colleagues. If you're planning on using a book club, we recommend that you keep the group to 3–5 individuals to make sure that you can build trust to fully discuss the sometimes difficult topics you'll encounter. Reading the book with a friend or in a small group may be beneficial, as you'll have the opportunity to discuss the content and scenarios and to practice some of the conversations in role-plays. If you're interested in forming a book club, check out the ideas and prompts we have outlined in the "Book Club Kit" at the end of the book (Appendix B).

You could also go through the book with a trainee or supervisee individually or in a small group, as the content can be very useful for those early in their career. We highly recommend that you complete the book first, or at a minimum, you remain at least one biweekly pair ahead of your trainees or supervisees so that you're familiar with the content and can be prepared to facilitate discussions and assign tasks. If you're taking this approach, you will find the "*Daily Ethics* Crosswalk" helpful (Appendix A). You can ask your trainees or supervisees to try to identify which standards they think are represented in the biweekly pair ethics topics and share why. Then you can check the "*Daily Ethics* Crosswalk" together and have a discussion.

Maybe you're the ethics compliance officer or point person for ethical dilemma support in your organization. If so, you could use the book to guide a review of your organization's practices. You could also use the book to present biweekly ethics topics for your team members by sharing a summary and some prompts in an email or group message. Again, you'll want to go through the book yourself first so that you're familiar with the topics and activities and can pick the right topics to introduce at the right time for your teams. You could even build continuing professional development activities using the content and offer continuing education units (CEUs). In that case, you'll also want to tie the topics to standards in the ethics code using the "*Daily Ethics* Crosswalk."

Whichever way you decide to go through the book, once you're familiar with the content in it, you may find that you use it as a resource in your clinical, supervisory, and leadership activities. One way you can stay connected with your daily practice is to flip to a random biweekly pair, review the content related to the topic and the notes you took when you first reviewed that biweekly pair, and then reflect on the tasks completed and those that you might identify as you revisit the content. Why? Well, that's how ethical dilemmas happen in real life. They don't come at you in any particular order, and in many instances, you might not see them coming. As you review the prompts, and even your notes from the first time through, you may find that your analytic skills and concept of the topic have evolved in meaningful ways. You may find that certain prompts resonate with you differently. If you're faced with an ethical dilemma, it may be helpful to flip to the related biweekly pair to help you navigate the current situation and develop a path forward to resolve the issue. This can also be very helpful if you're supporting someone who is worried about, or actively dealing with, an ethics issue.

Like any meaningful journey, you'll need to take breaks throughout. Breaks allow you to recharge, to engage in needed self-care so that you can continue with your daily practice on your journey. Breaks also allow you to reflect on how far you've come and survey where you are going. Don't skip these weeks. Embrace them and engage with the break activities. Believe us, you'll have earned them!

Biweekly Pairs

The ethics topics are presented in biweekly pairs to give you enough time to interact with the content in a meaningful way. You won't see specific standards from our profession's ethics codes, and that's very intentional on our part. Being familiar with the specific standards in the ethics codes is important! And likely, you took a graduate course in ethics that reviewed the standards, one by one. Hopefully, the relevant standards were also covered by your supervisor during your fieldwork experience hours. Early in the conceptualization of this book, we decided that it should stand on its own, separate from a formal code of ethics. From our perspective, doing so may provide you with a deeper connection to the topics. On this journey, we're looking to develop more than rule-following. We're looking to develop critical-thinking skills to help you navigate, interpret, and apply the rules or standards laid out in our profession's ethics codes. We have provided you with a crosswalk so that you can see the related standards in the current BACB ethics codes for each biweekly pair. You can access the "*Daily Ethics* Crosswalk" in Appendix A at the back of the book.

We also purposefully did not group the biweekly pairs based on the sections in the ethics codes or based on similarity. Why? Remember that we're hoping that you'll use this book every day in your practice. We don't know about you, but in our experience, ethics topics, issues, and dilemmas don't present themselves in an orderly manner. Life is just not like that. And again, your coursework likely, and rightly so, took a structured and orderly approach to teaching you about ethics. This book is meant to mimic the way we experience ethics topics in our daily professional lives. So, you may find a biweekly pair focusing on properly transitioning clients sandwiched between knowing your personal biases and multiple relationships. And we get it, that can feel a bit disorienting and frustrating, and we did that on purpose. Sorry; not sorry.

To balance that out, we'll provide some structure to the journey. The biweekly pairs follow a set configuration to lead you through the topic in a very intentional manner that moves from introducing the topic and some considerations, to prompting you to engage in reflection and action, and then facilitating some application and planning for continued work on the topic.

In the first week of a biweekly topic pair, you'll start off by reading some content to orient you to the topic. We start off this section with a quick check-in to give you a chance to reflect on the past week's topic and activities. We also start off with a check-in to give you a chance to take a deep breath before moving forward with the new topic. Then you'll read a brief section that describes the topic to be covered. Next, you'll be prompted to check in with your identified values (more on that in Chapter 3).

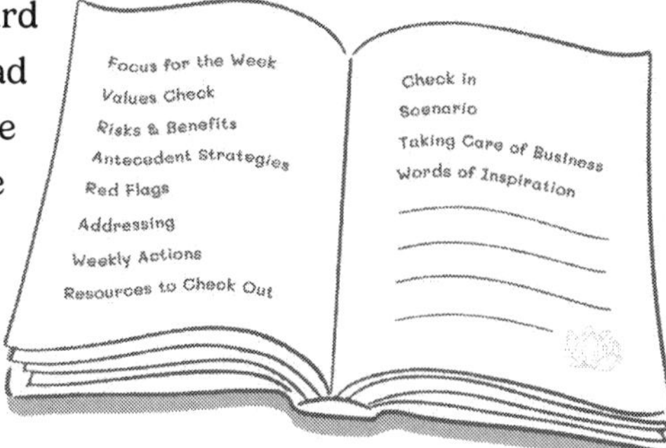

The next step on your journey with the topic is to think about the related risks and benefits. Then we'll present some information about possible antecedent strategies to help you actively engage with this topic to maximize the benefits and to minimize or avoid tangling with the ethics topic. Then we'll share some red flags that you can watch for to help detect an emerging or actual issue or dilemma. Finally, we'll share some tips for how to address a situation where this ethics topic comes up in a concerning or problematic way. Then you'll move into the daily, Monday through Friday, practice prompts for the first week of the topic.

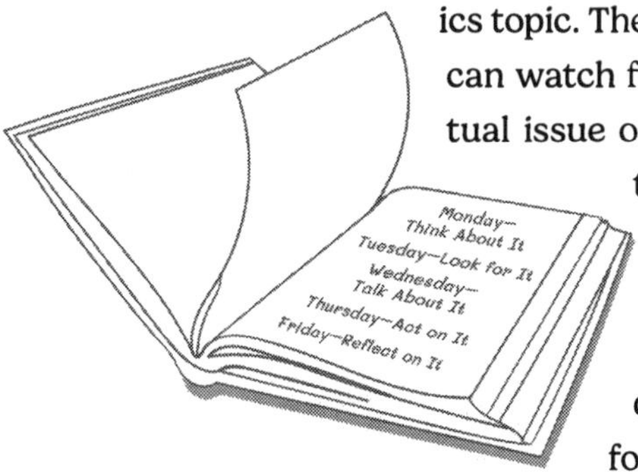

The first week on the topic ends with some additional resources for you to check out if you wish.

The second week on the topic starts off with, yeah, you guessed it, a check-in. This is effortful and not always pleasant, but it's good practice to build in pauses to take stock of how you're feeling and get ready to keep going. Then you'll find a scenario. The scenarios are pulled from our collective experiences in practice and represent amalgams of actual situations. You'll find a wide variety of "ethicality" represented across the scenarios. Don't be surprised if you read one and think to yourself, "There is NO ethical dilemma here." We did that intentionally to provide you with multiple exemplars of situations across the topics.

Remember, the goal of this journey is not to hone your skills of spotting ethics issues. Sure, we hope that you will develop that skill. But the goal of this journey is to combat our tendency to become complacent or avoidant. The goal is to activate your ethical muscle daily, to take an analytic approach to every situation, and to bring into focus the beneficial and risky ethics features of your daily interactions and activities. Some scenarios will contain clear issues, some may describe what feel like precursors to ethics issues, others may feel questionable, and still others may feel like shining examples of ethical behavior. Each scenario will end with a series of prompts for you to engage with the information presented and to reflect on your own behavior and practices. Then we'll ask you to connect with any tasks we've assigned. We'll wrap up the second week in the pair with some words of encouragement that'll hopefully serve to strengthen your daily practice.

Throughout the biweekly pairs, you'll find spots to jot down notes, thoughts, and reminders. If this is not enough space for you, we encourage you to get a notebook that can serve as a dedicated companion to this book. In the spaces provided, or in your notebook companion, we encourage you to describe your thoughts and feelings, summarize discussions you have, develop additional related scenarios to use in supervision, sketch or map out tasks and activities, and even just doodle. Remember, we're asking for you to practice, and that requires active engagement with the content. So, maybe you're a highlighter, an underliner, a circler, or an arrow drawer. It doesn't matter how you come to this journey and practice; just be actively engaged! You won't build ethical muscle or muscle memory by passively reading.

Most road maps have symbols to help you quickly orient yourself along your journey. On this journey, you'll find recurring icons to give you a heads-up about the content and activities as you engage in your daily practice. Here's a list of the icons you'll see and what they mean:

You'll see this icon a lot. Why? Because taking deep breaths is good for us! It reminds us to slow down. Deep breathing can help you "be where your feet are." In other words, taking a deep breath can help you connect to the present moment. Engaging in deep breathing can also help lower your blood pressure and reduce feelings of anxiety (Tavoian & Craighead, 2023).

This icon lets you know that the content or prompt might make you a bit uncomfortable. So, this is us telling you to do what you can to make sure you are comfy (e.g., take a few deep breaths, put on some calming music, make a cup of tea, get to a snuggly chair, go on a quiet walk) before moving forward. When you see this icon, it's a good idea to check in with yourself before proceeding. If life is a bit extra crunchy right now, make sure you are good to move forward or maybe decide to take a pause and reconnect when you've had a chance to refill your cup.

This icon is all about alerting you to the likelihood of upcoming challenging conversations. The purpose of this icon is to slow you down and remind you that we are likely to lean away from uncomfortable interactions, so we want to you prepare yourself to lean the other way; lean toward the learning opportunity, even if it feels challenging. Seeing this icon might prompt you to pause and call up strategies for preparing for a tricky discussion.

The self-reflect icon is meant to give you a heads-up that we are going to ask you to look inward, in an honest and thorough way. We know, from extensive personal experience, that most of us are pretty lousy at engaging in accurate self-reflection. When you see this icon, reconnect with your values and with your intention to create intentional daily ethical practice.

Now that you've got the lay of the land for your journey, let's get you prepared. The next chapter will walk you through some content and activities that are critical to successfully and fully engaging with your daily ethics practice. We'll walk you through topics like professional ethics versus personal morals, self-reflection, self-compassion, self-care, culturally responsive practice, values, and difficult discussions. So, please don't skip the next chapter. In fact, hang out there as long as you need, and feel free to return to specific topics anytime you need to during your journey.

Chapter 3

But First—Get Yourself Prepped

Now that you have a good feel for the what, why, and how of this book, let's get you prepped for your journey ahead. In this section we review some topics and have you engage in some critical activities that will set you up for success as you navigate the biweekly pairs and work to engage in daily ethics practice. First, we'll discuss some strategies that can help ease you into your commitment to make space for this daily practice, including some tips for engaging in self-reflection. Then we move into the topics of compassion for yourself and others, and strategies to care for yourself along this journey.

Some of the content we cover will likely evoke discomfort. We are all arriving here with a unique and complex intersecting identity that has been shaped by our experiences and beliefs. We want you to be as comfortable with this content as possible, and we don't want you to engage in escape behavior. A little purposeful avoidance can be healthy, but we want you to have the strategies to return to and engage with the content. The idea is not to be comfortable with discomfort, but to continue with your practice while you make space for that discomfort.

Then we'll cover the topics of implicit biases and cultural responsiveness so that you are prepared to lean into these topics when they come up in some of the biweekly pairs. After that, we'll focus on the importance of knowing your core principles or values and how being aware of yours can serve you when faced with ethical dilemmas and decision-making. Because ethical decision-making and addressing ethical dilemmas likely require us to have important conversations with others, we will spend some time covering the importance of identifying a potential tricky conversation, gauging other people's comfort level, and getting those critical conversations started. We even provide some scripting guidelines to get you going. Finally, we'll wrap up your preparation by highlighting how moving through this content will likely shine a light on some areas for professional growth. Ready? Let's get you prepped!

Prepping for Daily Practice

Remember how we talked about engaging in intentional practice to develop our moral or ethical muscle memory? For most of us, making decisions throughout the day happens with little purposeful thought (Bazerman & Gino, 2012). We tend to make decisions fairly quickly using rules of thumb, established practices, or simply what worked last time. And that is fine, for most of our decisions. But for more complex situations, like ethics issues, we will minimize risks, and even avoid them in some cases, if we increase our latency to responding by engaging in a more thoughtful and structured thought process. If we default to engaging in the response that is at strength, we may miss out on identifying nuances that can help us move forward in an ethical manner. We can pause and ask ourselves, *does my intuitive or at-strength reasoning and decision-making suffice here, or might I need something different, a little more structured?* (Bazerman & Gino, 2012). We also need to commit to challenging our tendency to think of ethics content as no more than right versus wrong or good versus bad and invite the messy, complicated reality that there are so many nuances to ethics content and dilemmas.

"Every day, those who desire to act ethically must remind themselves of their ethical aspirations and of the need to constantly keep their ethical antennae extended"

—(Drumwright et al., 2015, p. 438)

So how do we challenge our behavioral patterns that are currently at strength? Here are some strategies to help.

- Make a commitment to engaging in daily practice with ethics content. Set a goal! Say it out loud. Write it down. Tell a colleague. Better yet, commit to this practice together. Setting your intention doesn't mean that you'll nail it 100% of the time. But it does increase the likelihood that you'll engage in the practice more often than not, and that you'll return to it when your practice is interrupted by the inevitable illness, time-sensitive work tasks, unexpected out-of-town guests, unplanned staffing changes, and just the need to take a break.
- Add prompts in your environment to increase the likelihood of engaging in your daily practice. You might choose to make sure you leave this book in the center of your workspace at the end of each day. Maybe you can add a recurring event in your calendar for your daily ethics practice. You could add a sticky note or a funny meme to your computer screen or the wall by your desk to put in the work toward your daily ethics practice.
- Pair your practice with already recurring activities, especially if they are pleasant ones. For example, maybe you always have a morning tea, coffee, or smoothie. You could pair that with your ethics practice for the day. Or perhaps you take a regularly scheduled lunch break (if you don't, you should ☺) or a mid-afternoon stroll. You could engage in your daily ethics practice during those activities. The idea is that if you have some routines already built in your day, you can simply latch your daily ethics practice onto them.
- Track your progress. You could graph pages read or biweekly pairs completed. Maybe you make a bookmark with the "Biweekly Topics Checklist" on it and cross off each week as you get through the content. If you're going through this book with a colleague or in a group, maybe share your progress in the group to help keep each other motivated.
- Schedule access to reinforcers for completing targeted sections. For example, maybe you treat yourself to your favorite fancy coffee drink or a night at the movies for completing a biweekly pair.

Maybe one of these ideas catches your eye and you'd like to try it. Maybe none of these strategies resonate with you. The point is to find strategies that will increase the chances that you will actually pick up this book and engage with it more days than not. Is it reasonable to expect that you will do it every Monday through Friday without fail? Nope—it's not! And frankly, even though we are fond of the content, that doesn't sound very sustainable or enjoyable. It's reasonable that you'll have days where you can't, or don't want to, crack open this book and engage with the content. That's perfectly fine. We just want to make sure you set up some strategies to increase the chances that you'll return.

Prepping for Self-Reflection

Throughout this book, you'll be prompted to engage in self-reflection. What we mean by that is tacting your own past behavior, usually in relation to the behavior of others; identifying the likely relations and outcomes; and then evaluating that information against your values and goals and potentially identifying different ways of responding in the future that are better aligned with those values and goals. Self-reflection is important because only you have access to your learning history and private events. Also, it can be very difficult to observe ourselves objectively in the moment, so self-reflection lets us revisit situations after the fact so we can still learn from them. Reflecting on our past behavior can also allow us to evaluate our progress toward our goals.

Effective self-reflection also requires that we approach the activity with a relatively calm and clear mind. Let's be honest, a calm, clear mind is hard to come by in our profession. We are in constant motion, constantly observing and analyzing. Engaging in self-reflection can be tricky, especially if you have never tried it. It can make us feel uncomfortable, defensive even, which can result in us avoiding doing it. So, before we ask you to engage in self-reflection related to complicated ethics content, let's get ready by starting small. If you have experience with self-reflection or even meditation, you might only need to scan this section. If this content is new to you, uncomfortable for you, or evokes violent eye rolls, we just ask that you proceed with an open mind. Here are some strategies you can start playing around with. We give you the first week to practice self-reflecting and noticing too.

- Schedule time in your day or week to engage in self-reflection. It can be as short as 5 minutes. The idea is not to engage in lengthy sessions where you dissect past interactions, but to engage in regular and effective self-review of situations that can help you grow and make sound decisions.

- It is not necessary to always engage in self-reflection in the same environment. In fact, it is great to develop the ability to engage in self-reflection in a variety of settings. However, for your practice with the content in this book, it can be a good idea to identify a few reliably accessible, distraction-free places where you feel comfortable. For example, maybe there's a frequently empty office at work, a shady spot outside, a nearby park or café, or even your car. Just try to identify a few comfy spots.
- Make your comfy spot as appealing as possible. Maybe you can kick off your shoes, have a soft pillow to sit on, brew a cup of tea, or play some relaxing music.
- Practice sitting or standing in a comfortable position and simply being still. Don't worry about clearing your mind. Just put your phone away and be. If this feels difficult for you, try it for just 1 or 2 minutes. Do this several times a day, until it feels comfortable and you can just quietly be for several minutes.
- Try some deep breathing. You can try breathing in for the slow, steady count of four, hold for four, breathe out smoothly for a count of four, and hold for four. Do this three to five times. Once you're done, just observe how you feel.
- Start engaging in simple self-reflection by trying to recall a recent event or interaction. Start with mundane, everyday interactions or situations. Don't try to interpret things; simply replay the scene in your mind and tact your behavior and the behavior of others. Once you're comfortable with this, try increasing the complexity of the situation and how far back in time you're going.
- Now try adding some *wh*-questions to your reflection. Asking *wh*-questions can help you think more critically about past situations.

What did I think in the moment?
What do I think now, reflecting back?
What did/do I feel?
What do I think the other person thought/felt?

***What* was the outcome of that interaction or behavior for me/the other person?**

***Why* do I think I might have thought/felt that?**

***What* was my goal?**

***What* can I learn from that situation?**

***What* might I have done differently?**

***What* will I commit to doing differently?**

- You can try keeping a journal or a note on your phone or computer where you jot down situations that left you feeling a certain way. Maybe an interaction with a coworker left you feeling self-conscious, or a meeting with a trainee left you feeling joyful. Make note of the situation using just a few descriptive words, and then review those notes in your next scheduled self-reflection opportunity.

Because sometimes self-reflecting can help us identify mistakes we made and things we would like to correct or do better. And do be sure to be kind to yourself. You'll learn more about that in the next section.

Prepping for Compassion, Self-Compassion, and Self-Care

The science of ABA is founded in compassion (Penney et al., 2023). Compassion might be one of those words that you use but have never really had to define. One widely accepted definition is "A deep awareness of the suffering of another coupled with the wish to relieve it" (Gilbert, 2009, p. 13). Strauss et al. (2016) reviewed and synthesized existing definitions of compassion and offered that it involves the following five components:

1. recognizing suffering in another
2. understanding that suffering is universal
3. feeling empathy (i.e., connecting with the other's suffering)
4. tolerating the discomfort we feel in response to the person's suffering
5. acting to minimize or alleviate the suffering

Recently, behavior analysts have proposed a behavior-analytic definition of compassion as engaging in empathetic perspective-taking that moves us to act to alleviate the short- and long-term suffering of another by minimizing or removing aversive stimuli (LeBlanc, Taylor, & Marchese, 2020; Rodriguez et al., 2023; Taylor et al., 2019).

Along your journey through this book, you will likely become frustrated with the behavior of others, and we ask that you pause and invite compassion. Let your frustration be a signal to pause, take a breath, and ask yourself if their actions could at all be related to them contacting aversive stimuli. Ask yourself, *could they be suffering*? Moving through your daily ethics practice from a place of compassion will invite you to further investigate the environmental contributors of tricky or problematic ethics scenarios and dilemmas, instead of engaging in the knee-jerk reaction of blaming the person.

This also means that you should lean into practicing self-compassion; because, let's face it, none of us are perfect. Sometimes the person who frustrates you will be yourself. Neff and Knox (2020) beautifully describe self-compassion as meeting life (most importantly, our missteps and mistakes) with "an open-hearted stance in which the boundaries of self and other are softened so that all human beings are considered worthy of compassion, including oneself" (p. 4664). So, we invite you to meet your past, current, and future mistakes with an abundance of self-compassion. Don't get caught up in beating yourself up with unkind self-talk. Lean into what the situation can teach you. In fact, you can even stop and ask yourself, *what lesson is this mistake offering to teach me?*

Engaging in self-compassion also makes space for actively practicing self-care. Self-care activities are those that proactively support our health (physical and mental), reactively reduce our stress, and help us return to a state of well-being. Self-care involves managing work and life stressors and accessing resources and supports to maintain or return to a state of well-being. We cannot prescribe specific self-care activities or processes to you, as you are a beautifully unique and complex person. You take care of a lot of people, and leaning into daily ethics practice can be challenging, so we want to be sure that we are also sending a loud and clear message that you are important, and you need to take good care of yourself. So, you will have to discover those for yourself. Across the weeks, you'll find several prompts and resources to engage in self-care and wellness activities. You might start by learning more about your level of self-compassion with a simple online assessment (Neff, n.d.).

Prepping for Implicit Biases and Cultural Responsiveness

Let's start with what we mean by implicit bias and cultural responsiveness. When we talk about implicit bias, we are referring to the automatic reactions we have to people and situations that affect how we respond, without us being aware of the impact it is having on us (Project Implicit, n.d.). When we refer to cultural responsiveness, we mean the act of responding to the cultural variables that everyone brings and engaging with them in a way that values the uniqueness of each individual (American Speech-Language-Hearing Association, n.d.). You may have also come across the terms *cultural competence* and *cultural humility* in your practice. These terms are all related, and for the purpose of this book we will be using *cultural responsiveness*. We'll define these terms more in Weeks 23 & 24.

In our practice as behavior analysts, we have an ethical obligation to our clients, supervisees, trainees, and others to be aware of and address our implicit biases and engage in culturally responsive practice. If we do not do this, we run the risk of engaging in behavior that might minimize the perspective, experience, and values of those we are working with, and we may be providing services and supervision that may not be effective or may even be harmful.

It might be the case that you are just starting your journey of thinking about, identifying, and discussing implicit biases and working toward cultural responsiveness. Or maybe you've been doing this work for a while and are very familiar with identifying your implicit biases, responding to them, and engaging in culturally responsive behaviors. Maybe you find yourself somewhere in between. Wherever you are in your journey, it's okay! As you work through the biweekly pairs, you'll find prompts to think about and address your implicit biases and reflect on your culturally responsive practices as a behavior analyst in relation to the specific ethics topics. Not only will this topic come up in multiple biweekly pairs, but we'll also dive a bit deeper on the topic in Weeks 15 & 16 and 23 & 24. But we want you to be well prepared to hit the ground running.

We want to take a moment for some real talk. This work might be uncomfortable. You're likely to identify areas where you might be doing really well and areas where you have room for growth. You might do this work reflecting on your own or you might do this with a colleague or supervisor. To help you prepare for these conversations and reflection, here are some tips:

- Access and review resources to help you self-evaluate where you are in this learning process.
  - You might consider starting with Project Implicit's Implicit Association Tests (IATs) (Project Implicit, 2011) to start to identify your implicit biases. (We can share that we were surprised with some of our results when completing these tests. Remember that we all have biases, and the first step is just to be aware of what they are.)
  - The article by Beaulieu and Jimenez-Gomez (2022) provides a nice description of the relevant terminology and provides some recommendations for self-assessment.
  - The Fong et al. (2017) article is a great place to start to get some background on the importance of this work, some challenges that are faced, and some recommendations.
  - Looking for resources to inform your supervisory practices? Mathur and Rodriguez (2022) propose a cultural responsiveness curriculum and provide a checklist to assess competency.
  - Neff (n.d.) provides great resources for engaging in self-compassion, including guided practice and an evaluation of your self-compassion.
  - Want to learn more about the terminology? The article by Tervalon and Murray-García (1998) provides recommendations for moving to cultural humility in terminology rather than cultural competence.
  - If you find that you are ready to engage in self-reflection, check out this article by Wright (2019).
- Complete a self-evaluation to identify where you are in your journey to culturally responsive behaviors.

- Take some time to reflect on your self-evaluation. Where do you think you might need to focus some more time, self-reflection, and learning? Remember those self-reflection tips you just reviewed? Those are going to come in handy here!
- Think back on your supervision and training. Did you have conversations with your supervisors about identifying your implicit biases and what to do after identifying them? Did you discuss culturally responsive practices during your supervision and training? If you are a supervisor, do you address these topics with your trainees and supervisees?
- Prepare to engage in perspective-taking as you think about and address implicit biases and engage in culturally responsive behavior. Is it easy for you to invite and consider the perspectives of others?
- Spend some time thinking of a few colleagues, supervisors, mentors, and friends who you can talk to about this content. Having a safe space to reflect on and talk about these topics can be very helpful.

Alright, take a deep breath. Be kind to yourself. This will always be an area of growth for all of us. Acknowledging and addressing our implicit biases and cultural responsiveness is a moving target that we will always be working toward. And we all have implicit biases. We can't change that we have them, but what we can do is acknowledge their existence and work to change how we react (or don't react) to them.

Prepping for Identifying and Using Your Values

Each day, we wake up and go about our day in a specific way. There are so many influences and contingencies that shape our behaviors, but an interesting variable to consider is our values. Values are like the theme song playing in the background of our lives or a common thread woven through our days that ties together our decision-making, priorities, and behavior. Values reflect a theme of the things we would like to get out of life (Fiebig et al., 2020), as opposed to goals that can be achieved and checked off. Our values influence how we show up in the world each day, what we lean into or away from, what and who we make time for, and how we engage with our world.

Now, we're not here to provide you with values to adopt or to preach our values. Instead, our aim is to provide you with a framework and space to self-reflect on values that serve as your true north. We want you to know your values today, how they've changed over time, and how they inform what you do and what you *don't* do.

Our values are unique and authentic to our beliefs, history, perspective, and priorities. They will likely shift through time and experience. What was most important to you in your youth may not have the same importance to you in your adulthood, but there are likely common threads that are just as important, if not more so. No matter what they are, if you're aware of them or not, or how many times they have shifted over the years, they play a large part in how you live your life. So, if they're that important, it probably makes sense to figure out what they are!

Knowing Your Values

There's a difference between knowing your values and living them, but if you're beginning to think about what your values are, a good place to start is reflecting on your life in a mirror. When you look at your reflection, consider the questions below. What themes do you notice in how you show up each day?

We have a few approaches that you might mix and match as you start to identify or refine your values.

First up, consider the following questions and jot down what comes to mind:

- Where are you when you feel most alive?
- Who do you strive to be more like, and why?
- What brings you energy and joy?
- What situations do you respond to and always make time for?
- What do you most appreciate in others?
- When do you feel most proud? Why?
- Who do you most enjoy spending your time with, and what qualities do they possess that you admire and appreciate?
- What are you doing when you are content and at peace?

____

____

It's equally important to reflect and identify the relationships and situations that you respond away from; those that are likely misaligned with your values. The following questions will not necessarily define your values but may be helpful to consider when identifying what you do not value.

- When do you feel most anxious?
- What situations do you not want to be in, or do you move away from?
- Who are those people you do not respect or want to be around? What behaviors do they engage in to make you feel that way?
- When do you feel uneasy or experience a pit in your stomach that just doesn't feel right?
- What activities or tasks do you procrastinate or avoid?

Another approach comes from Cozma (2023), and it suggests that you:

1. Reflect on what's important to you.
    a. Pick your top three values.
    b. Rank them.
2. Define your values, as your definition of a value like independence could be very different from someone else's.
3. Use your values (more on that coming up).

There are other resources and activities that you may find valuable to help you in exploring and discovering your values. You can find online lists of values words to review and reflect on. Personal Values (n.d.) offers a quick and easy test to identify your values. We tested it out, and some of our values include family, love, curiosity, health, challenge, adventure, peace, and authenticity!

Take some time to review your answers and notes. Look for common themes that can be summarized as your values anthems. Then, think about how your daily behavior and choices align with or against your values. Be patient with yourself. Did you engage and stay true to how you want to show up today? What decisions will you make tomorrow and the next day to further align with your values?

Living Your Values

The only thing more important than knowing our values is living them. Once we know them, how do we behave in a way that honors and reflects their evolution and impact on our lives over time? It can be difficult enough to remember our values—let alone live them—in times of conflict and stress when we're likely to move away from discomfort. It's likely easier to show up and bring our values to life when someone is asking for advice or when things are going well, versus living our values when we lose our job, our spouse tells us they want a divorce, or we're experiencing a life-threatening health condition. These challenging events can also present us with the greatest opportunity to reflect and know what our values truly are—who we are and who we want to become—and the daily behaviors we engage in to get us there.

We have all lived through difficulties. Adversity tends to strip away any façade that we work hard to create and leaves us with nothing but ourselves, our beliefs, and our values. What have you learned from and about yourself through your difficult and challenging experiences? We cannot always change what happens, but we can control how we respond and learn as we move forward to become our best selves.

There are situations that have, or will, tested us and placed our values in direct conflict with our ethics codes or the policies and procedures of our organizations. Maybe a couple you work with asks you to babysit for them so they can go on their first date night in years. They tell you you're the only one they trust. You're conflicted between your values of compassion, kindness, and service to others and your ethical obligation to avoid a multiple relationship. Maybe you're offered a gift of significant value that puts you in a vulnerable spot. You know that your ethics code says you should not accept it, but your values include being respectful, grateful, and culturally responsive.

As you work through the biweekly pairs, we'll highlight considerations and perspectives to navigate these complexities in ways that won't end up putting you in direct violation of your roles and responsibilities or who you are as an individual. You'll have to do the work of learning to continually connect with your values in the context of ethics content and dilemmas. Remember that not everything that goes against your values is unethical and not everything that aligns with your values is ethical. The key is being aware of your values and gaining tools that you can use to avoid, mitigate, and address risks. We need both our values and tools to balance us as we walk what sometimes feels like a tightrope.

Living Your Values in This Book and Beyond

Your values, and the foundational values of the profession, are core structural components within your ethical framework, practices, and decision-making. Knowing your values sets you up to better live your values, especially under challenging circumstances. As you use this resource to strengthen your ethical decision-making, ground your awareness and ethical responsibility in the values that you have defined for yourself. It's likely that we already share many of the same values because you are reading these words. We probably share in the value of self-improvement, doing what's right, and helping others. Preparing for and navigating through ethical considerations and decision-making is not easy. But, knowing how you want to show up and taking the steps to get there will ensure that you have built a strong foundation to stand on. We each have a limited amount of energy and effort that we can allocate each day. Making intentional decisions to live our lives according to the values that we believe is a courageous step. We applaud you for leaning in and are grateful to be a part of your journey.

Prepping for Uncomfortable Conversations

Talking about ethics content is so important, as hearing the perspectives of others increases our understanding and ability to engage in moral imagination (i.e., the ability to imagine a variety of other ways an ethics issue might play out). It also gives us a chance to state our thoughts out loud, which probably helps us better engage with our thoughts and ideas as a listener and facilitates self-evaluation. Throughout this book, we ask you to have a lot of conversations. In the first week of each biweekly pair, we ask you to talk about the topic with peers, supervisors, and your supervisees and trainees. And we want to be honest with you: It is very likely that actively engaging with the book content will result in some uncomfortable conversations. You may need to have a difficult conversation because you need to take ownership for something or ask for support. You might need to talk about your concerns about someone's behavior. You might need to talk to your leader about concerns or ideas related to ethics issues or needs in the workplace. We are not sorry about this eventuality. Yeah, these conversations can make us feel anxious; we can worry about how we will be perceived or if we might hurt someone's feelings. They are also necessary if we want to grow and help those around us grow. We do want you to be prepared for these conversations, so here are a few tips:

- Engage in professional development activities related to feedback and difficult conversations. Look for continuing education webinars, workshops, and talks. Consider reading books on this topic alone or in a book club. Some great options include:
    - *Crucial Conversations* (Grenny et al., 2022)
    - *Difficult Conversations* (Stone et al., 2023)
    - *Nonviolent Communication* (Rosenberg, 2015)
- Think about your communication style and how you typically respond to tricky or uncomfortable conversations. If you are someone who tends to avoid them at all costs, then you might commit to leaning into having them. If you are someone who tends to be overly direct, then you might consider working on perspective-taking and how to approach these conversations more collaboratively.
- Prepare by tacting that the conversation is likely to make you or others uncomfortable so that you can plan an appropriate approach to the topic. Identify clear, shared objectives and your desired outcome for the conversation. For example, are you asking for support, fixing a mistake you made, expressing your concerns, or offering up ideas for improvement? It can be helpful to write out some scripts (especially to just get the conversation going), predict responses (yours and theirs), and then practice.

- Be thoughtful about the time and place of the conversation to increase the likelihood of success and minimize placing the other person on the defensive. Most of these conversations should happen in a private, distraction-free, safe space at a time when you and the other person can focus on the topic at hand.
- Open the conversation by stating what you want to talk about and why (framed in a shared purpose or goal), using neutral and nonjudgmental language. Be clear and avoid language that is ambiguous or requires interpretation.
- Engage in active listening and attend to your partner's behavior so that you can adjust your approach if needed. Ask open-ended questions about their perspective, paraphrase their responses back to them, and ask if you got your summary right.
- Be compassionate by demonstrating empathy, respect, and validation throughout the conversation. Continuously assess the other person's comfort level by watching for signs of confusion, distress, or defensiveness. If you notice discomfort, consider taking a step back, rephrasing your statements, allowing them time to process the information, or asking how you can support them in moving the conversation forward.
- If a solution to an issue is needed, ask to work collaboratively to identify one. Offer and solicit suggestions and collaborate with the other person to find common ground and develop a plan for moving forward.

Difficult conversations can be challenging, for sure. Most of us never received any formal training on how to initiate or navigate these tricky conversations, and for many of us, we have plenty of models of ineffective or even harmful difficult conversations. But uncomfy conversations are necessary for the growth and improvement of our skills and practices, particularly when it comes to ethics. The more we lean into talking about ethics content and having tricky conversations, the better we will get at having these discussions and the less scary they will become for everyone. By approaching these conversations with empathy, active listening, and a focus on collaboration, you'll be able to create a more supportive and comfortable environment for the conversation. If you're a supervisor, consider explicitly teaching supervisees and trainees skills related to difficult conversations; that's how we pay it forward.

Prepping for Identifying Professional Growth Opportunities

In addition to increasing your fluency in thinking about ethics on the daily, it's important that you tend to your professional growth. Many behavior analysts are required to obtain CEUs by their certifying and/or licensing organizations. If you're not planful with identifying topics and skills that need attention for maintenance or growth, you run the risk of viewing these professional development requirements as a chore; as something to be checked off, and often at the last minute. Approaching your professional growth opportunities in this manner is haphazard and may result in accessing less-than-optimal learning opportunities in the interest of time and convenience.

But it doesn't have to be like that. You can take a purposeful and active role tending to your scope of competence and professional growth. As you move through the content in the book, you'll likely identify areas where you might benefit from a refresher to ensure your skills are up to date with current developments and standard practices. You might identify areas where knowledge and skills are underdeveloped or lacking. Don't be surprised. None of us have had the perfect education, training, or work experiences. We all have areas for improvement. So, be kind to yourself as you identify skills you need, or you'd like, to develop; articles you'd like to read; and areas that could benefit from coaching or mentorship. We suggest you keep a running list of such ideas and fit them into your ongoing professional development plan.

Use those ideas to set goals for yourself and help you decide how you're going to allocate your resources (e.g., professional development funds, needed CEUs, time and energy). You might check out the resources on professional development that the Association of Professional Behavior Analysts (APBA, n.d.) provides for its members. Specifically, you might find the documents titled "Professional Development Strategies for Assessing Needs & Creating Your Plan" and "Professional Development Needs Assessment" helpful in taking an active approach to your professional development. Think of your professional knowledge and skills as a garden that you need to continually tend so that it can grow healthy and strong. We are excited for you to watch how your garden grows, blooms, and evolves over time.

Prepping for a Little Fun

We've been honest that you're likely to feel some discomfort as you work through the biweekly pairs of ethics topics. We also discussed the importance of engaging in self-care activities earlier in the chapter, and hopefully, you started to get some ideas. One way that all three of us engage in self-care is listening to music! Aaaannnnd, we've created a playlist to share with you! Welcome! ♫ We hope these songs energize, empower, and uplift you to help keep you motivated. Maybe they even make you want to dance a bit! Know that they are in no particular order, and some of them have been selected because we just plain enjoy them. If they aren't all for you—that's okay! Hit that skip button. Maybe our playlist will motivate you to create your own that can keep you energized and motivated as you work through the book. Happy listening!

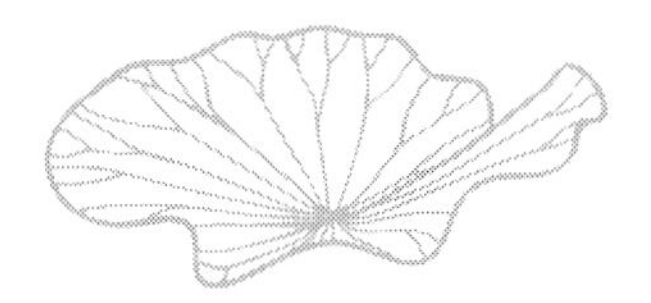

Chapter 4

Week by Week—Get Going!

Alright, alright, alright! It's time to get started!!! Wait a minute ... Did you read Chapters 2 and 3? If not, we highly recommend that you go back and do that. It's okay; we'll wait.

Okay, all ready to go? Remember, you get to control the pace. It's okay to skip a biweekly pair and come back to it. It's fine to take an additional rest or reconnect week to give yourself time to recharge or get things done. It's perfectly alright to close the book and put it away for a little while, although we hope you don't, as the idea is to develop your daily practice of interacting with ethics topics. To help keep you organized, you can use the week-by-week list below to check off the weeks as you complete them, highlight weeks you might skip so you don't forget to go back and do them, make notes about weeks you complete but might want to go back to later, and celebrate all your checked-off weeks as you progress through the book.

Biweekly Topics Checklist

- ☐ Weeks 1 & 2—Prepping for Your Daily Ethics Practice
- ☐ Weeks 3 & 4—Knowing & Evaluating Your Scope of Competence
- ☐ Weeks 5 & 6—Building Collaborative Foundations With Caregivers
- ☐ Weeks 7 & 8—Knowing Your Supervisory Responsibilities for a Strong Start
- ☐ Weeks 9 & 10—Resting & Reconnecting
- ☐ Weeks 11 & 12—Protecting Confidentiality
- ☐ Weeks 13 & 14—Tending to Your Supervisory Practices
- ☐ Weeks 15 & 16—Facilitating Your Awareness of Biases
- ☐ Weeks 17 & 18—Resting & Reconnecting
- ☐ Weeks 19 & 20—Addressing Your Own Possible Misconduct & Possible Misconduct of Others
- ☐ Weeks 21 & 22—Focusing on Compassion in the Therapeutic Caregiver Relationship
- ☐ Weeks 23 & 24—Connecting With Culturally Responsive Practices
- ☐ Weeks 25, 26, & 27—Resting & Reconnecting
- ☐ Weeks 28 & 29—Assigning Tasks to Supervisees & Trainees
- ☐ Weeks 30 & 31—Addressing Professional Boundaries & Multiple Relationships
- ☐ Weeks 32 & 33—Locking Down Documentation
- ☐ Weeks 34 & 35—Resting & Reconnecting
- ☐ Weeks 36 & 37—Transitioning Clients Responsibly
- ☐ Weeks 38 & 39—Watching for Power Imbalances & Exploitation
- ☐ Weeks 40 & 41—Ensuring Appropriate Consent & Assent Practices
- ☐ Weeks 42 & 43—Resting & Reconnecting
- ☐ Weeks 44 & 45—Discharging or Discontinuing Clients Professionally
- ☐ Weeks 46 & 47—Addressing Romantic or Sexual Multiple Relationships
- ☐ Weeks 48 & 49—Dealing With Interruptions, Transitions, & Terminations of Supervisory Relationships
- ☐ Week 50—Catching Your Breath
- ☐ Week 51—Catching Up
- ☐ Week 52—Reflecting & Planning

Weeks 1 & 2—Prepping for Your Daily Ethics Practice

Week 1

Check in—Welcome to Week 1! We haven't even tucked into any specific ethics content, and you have already worked hard up to this point, so these first 2 weeks are about building a foundation for your practice by getting you familiar with the framework. Each of the biweekly pairs will start out with a check-in just to see how you are doing as you move through the content and work on your practice. Since this is the first week (GAH! So exciting!) we will focus on checking in on how you are feeling after going through the getting prepped content. We hope that you have a solid understanding of why it is so important to build your fluency with ethics content through purposeful daily practice. We also hope that you found some useful information and leaned into the topics of self-reflection, compassion and self-care, biases, cultural responsiveness, connecting to your values, difficult conversations, and attending to professional growth opportunities.

Focus for the Week—This is where we will introduce the specific ethics topic in week one of the biweekly pairs. For our first week together, we aren't focusing on any specific ethics topic. Instead, we are going to just spend some time on the first two overarching components of your daily practice, and those are thinking about (via reflecting) and noticing elements related to ethics in your professional behavior and in the workplace. These are essential parts of your practice because when we keep ethics at the forefront of all our clinical, supervisory, and business activities, we increase the likelihood that we will behave in ways that are aligned with our values and our professional ethics and that we will catch possible or developing issues early before they cause harm.

Values Check—In Chapter 3, we talked about how important it is to know your values. We provided some suggestions for how to identify yours so that you can use them in your professional work to move toward values-aligned behavior. Reminding yourself to think about and to notice ethics-related content throughout your professional day can be tiring and uncomfortable. Look back at your values. Can you identify one or two that seem to be directly linked to doing the right thing, even when it is challenging or scary? Write the specific one or two values in the space provided below. You will want to come back to these value words when things get tough.

Risks & Benefits—In the biweekly pairs, you'll read about the risks and benefits associated with the ethics topic. For this week, let's just talk about the risks and benefits related to thinking about and noticing professional ethics-related situations around you. The main risks of not thinking about and noticing ethics-related issues in your professional practice are no brainers: violating ethical obligations and resulting in disciplinary actions, harming clients and other people, impacting a business, and damaging the profession. But other risks are more subtle, like the slow deterioration of ethical practices over time and the shift toward a work culture that breeds coercive and toxic behavior.

The benefits of thinking about and noticing ethics-related elements are obviously that we avoid or minimize those awful risks. Clearly, thinking about and noticing ethics allows us to identify potential and actual risks or harms and take swift action. But if we look deeper, another benefit is that we create clinical, supervisory, and business practices that actively center critical ethics principles like safety, beneficence (i.e., doing good above all else), accountability, and autonomy and respect for others. That sounds like a place we'd like to work!

Antecedent Strategies—In your biweekly pairs, we will highlight some antecedent strategies related to the specific ethics topic. Related to thinking about and noticing ethics, you are already well on your way to implementing what may be the most important antecedent strategy, and that is starting with YOU. You have unique information about your history, are the only one to have access to your private events, and are the most effective behavior change agent for your own behavior, so committing to developing your fluency with thinking about and noticing ethics in the professional context is the fastest way to head off ethics issues.

So, making a specific commitment, engaging with this book, and adding prompts to your environment are great antecedent strategies. In the workplace, some proactive steps you can take are to share your thoughts about ethics related to specific issues or practices, prompting others (especially your supervisees and trainees) to think about, notice, and even build these practices into agendas and early steps in developing policies or processes.

Red Flags—This section will focus on discussing red flags that could be markers of potential or actual unethical behavior. For this first week, we will instead focus on red flags that could indicate barriers to proactively thinking about and noticing ethics-related elements. For most of us, our coursework and other structured activities related to ethics have likely focused on identifying when someone else is engaging in concerning behavior and identifying specific ethics standards at play. And those are excellent activities. However, if that was the bulk of your ethics training, that might be a red flag that what is at strength is (1) to wait until an ethical dilemma presents itself to think about ethics, and (2) to singularly focus on other people's behavior.

Another red flag is there could be risks stemming from barriers that prevent proactively thinking about and noticing ethics including how ethics content was framed in the past and how it is framed now. If your experience with ethics content is that it is boring, or scary, you are more likely to avoid regularly and proactively engaging with the topic. If you have been, or are, in a culture that breeds guilt and shame and punishes mistakes, then that could also be a red flag that you and others will avoid proactively thinking about and looking for ethics-related issues or opportunities. Two surefire red flags are if leaders approach ethics topics from a place of righteous indignation or if they don't talk about ethics at all.

Addressing—Typically, we will talk about strategies for how to actively address the specific ethics topic or concern you might have about your own or other's behavior. For this week, let's just start by reminding you that you are already well on your way to increasing your skills related to thinking about noticing ethics in your behavior and in the professional workplace. So, for this week, let's move toward addressing, with yourself, if you notice any barriers to actively thinking about and noticing ethics-related elements in the workplace, but more specifically, in your behavior. To get started, maybe have a little conversation with yourself linking this practice to your values and making a commitment to your daily practice.

You might notice that we keep referring to "professional ethics" or "ethics in the professional context." Sorry if that is annoying, but remember, there are important differences between our personal morals and professional ethics. This book focuses on honing your skills related to professional ethics in the work context, and that is very different than exploring and developing your personal moral or ethical practices.

Weekly Actions—In the biweekly pairs, we have provided a different focus for each of the days of the week in the first week. You'll start the week thinking about the topic and then noticing related things in your professional practice and in the workplace. Midweek you'll work on having discussions about the topic. Then you'll end the week by taking action, making a to-do list for tasks that will take more time, and then reflecting back on your week and forward into your ongoing practice to commit to continued focus on the ethics topic. This week we will introduce you to that framework as you stick with practicing thinking about and noticing.

MONDAY THINK ABOUT IT

- Spend some time thinking about your thoughts and feelings about professional ethics. Explore what thoughts and feelings come up in relation to the word ethics. If you are a visual person, write the word "ethics" on piece of paper or look at an ethics book or professional code of ethics document. If you are more of an auditory person, maybe say the work "ethics" out loud several times or play a podcast episode focusing on ethics in the background. Now just notice your thoughts and feelings. Do you feel slightly anxious, bored, excited? Are you hyperfocused, or do you find your thoughts wandering to what you should make for dinner and if the new episode of your favorite show will drop this Friday? Do you have butterflies in your stomach, are your palms sweating, or are you feeling sleepy? Maybe write down your thoughts and feelings, doodle or sketch things that illustrate your thoughts and feelings, or even use a voice memo to capture what is going on.

- Think back to your coursework and supervised fieldwork experience. Were you trained to think about ethics proactively or reactively (e.g., in response to concerns about someone's behavior)? Were you expressly taught to consider ethics in relation to all your professional activities? How about in relation to your own behavior? Were you expressly taught how to engage in active noticing for the purpose of catching possible issues early?
- Take some time to notice the first things you think about when you focus on ethics. Maybe the first things that come to mind are professional ethics codes, or a book or article, or a person's name. Maybe it's examples of other people's unethical decision-making and behavior. Notice where your decision-making and behavior show up on the list of things that come to mind.
- Think about your thinking related to ethics. We know, we know. That sounds super meta. Here is what we mean: When you think about something like an ethical dilemma, what is your thought process? Do you lean toward rigid, or black and white thinking? Do you focus immediately on laying out your rationale for why the person did a bad thing and what a bad person they are? Or do you find your thought process a bit more flexible, driven by looking at the situation from multiple vantage points and focusing on understanding the risks and possible causes over determining that an ethics violation occurred and what the consequences for the person should be? No judgment here. We just want you to observe and get familiar with your baseline approach to thinking about ethics topics.

- Evaluate how frequently you think about ethics. Look back at the past 6 months and notice how frequently you actively thought about ethics, and try to be specific. Of the times you thought about ethics, what percentage of the time were your thoughts evoked by (1) an unfolding ethical dilemma, (2) a prompt or question from someone else, (3) some other stimuli in the environment (e.g., social media post or article in the BACB's newsletter)? Since getting started with this book, has the frequency changed? If so, how do you feel about that?
- Notice how you feel after setting aside time to think about these topics. Are you tired? Sad? Frustrated? Excited?
- Jot down notes and thoughts:

TUESDAY—LOOK FOR IT

- Get started building your noticing behavior by scanning your own professional behavior. Where does ethics content show up? How often do you engage with books and articles on the topic? At what points in your clinical practice do you engage in specific behavior that functions to ensure ethical practices and evaluate for possible risks or harms related to unethical practices? How about in your supervisory or business practices?
- Have you ever noticed someone else's behavior that made you concerned about ethicality? Have you ever noticed a policy or standard practice in the workplace that worried you? Have others pointed these things out to you? Have you, or has someone else, ever noticed something concerning with your practices?
- In your workplace, are ethical and unethical practices clearly defined to facilitate noticing positive examples and issues? Is there a well laid out process for what to do if someone notices an area for improvement or has a concern? Does everyone know about the process and feel comfortable following it?
- Jot down examples:

WEDNESDAY—TALK ABOUT IT

- Talk to a supervisor or mentor about this book. Consider asking them how they think about ethics and what they do to facilitate actively noticing. Talk about the benefits, and drawbacks, of working to think about ethics daily. Explore the risks of becoming hypervigilant, seeing an ethics issue around every corner, and being what could amount to the ethics police (i.e., always tacting for people what they are doing wrong). Compare this to using a daily focus on ethics to enhance therapeutic relationships, safety, autonomy, compassion, and respect.
- Maybe chat with your supervisor about the current culture in the workplace related to ethics. Maybe the culture is great! If so, that is wonderful, and maybe you can focus on strategies to strengthen things by prompting more frequent, proactive talk about ethics. If the culture is susceptible to improvements, well, increasing the frequency of just talking about ethics broadly is a great stepping stone along the path to making positive improvements.
- Introduce the concept of thinking about and noticing ethics elements in our own behavior and in the workplace. You can even use some of the prompts from Monday to facilitate this conversation.
- Jot down notes and thoughts from your conversations:

THURSDAY—ACT ON IT

- You might feel like, what is there to act on? All I've been doing this week is thinking, thinking about thinking, and thinking about noticing! For now, we'll ask that you look back over the "Prepping for Daily Practice" section, review any notes you made, and take small actions to increase the chances that you will follow through with your commitment to think about ethics each day.
- Jot down thoughts and tasks:

FRIDAY—REFLECT ON IT

- Okay, you're almost done with the first week of this biweekly pair. Get to your comfy place. Deep breath. Take some time to reflect on how you felt this week. How did it feel to actively think about ethics each day? How did it feel to lean into noticing things in your behavior and in the workplace? How about talking to others; did that feel weird? If you did manage some conversation, reflect on how you felt before, at the start, during, and after. How do you think your communication partners felt?

- And listen up, don't be mad at yourself, past professors or supervisors, or current colleagues or supervisors. We are all a function of our learning histories and environments. Cultivate compassion for yourself and others as you develop your daily practice.
- Jot down something new you learned or something that surprised you:

Wrap up—Now you've had a glimpse of the structure for the first week in the biweekly pairs. It's okay if it feels awkward or uncomfortable. Some or all of these practices might be new to you, or at least they might feel new as applied to the topic of ethics. You're doing great! As you move through the first one or two biweekly pairs, we hope that these activities start to feel comfier. Typically, the first week of a pair will end with a section called "Resources to Check Out" that you can use to dig into the ethics topic, but we don't have that section for this week. You might want to take a few minutes to flip back to the resources we shared in Chapter 3 that focused on compassion for others and yourself and identifying your values.

In the biweekly pairs, the second week is really about application and time to get tasks done. It will start with a quick check-in. Why? Because we recognize that you, the reader, are a real person with all the complexities that come with humanity. And, because we care about you. Is that weird? We don't think so! Sure, we probably have never met, but we three want nothing more than for you to feel supported, encouraged, and appreciated on your journey to build daily ethics practices.

The next thing you'll find is a scenario and then some prompts to invite you to actively engage with the scenario. The second week of the pairs includes a brief reminder for outlining your current and future tasks and then ends with some words of inspiration for you. When you move into Week 2, you won't see that structure. Week 2 has the same structure and sections as Week 1 but focuses on having you practice getting ready to talk about and take action. See ya next week!

Week 2

Check in—Welcome back! Hopefully your noggin has recovered from all the deep thinking we asked you to do last week. This week we will continue to build a strong foundation for moving through the biweekly prompts included in the first week of the pair.

Focus for the Week—Last week you worked on thinking about and noticing ethics content and elements. This week we'll have you shift to getting ready to talk about and take action on ethics content. This week should hopefully feel easier and simpler than last week since you sort of know what to expect. We also purposefully kept this week a bit lighter so that you are not exhausted going into the biweekly ethics pairs.

Values Check—Check in with your values. Which ones feel most related to talking about ethics, particularly leaning into uncomfortable conversations, and taking action to improve your practice and the practice of others? Write the specific one or two values in the space provided below.

Risks & Benefits—This might feel like a "duh" moment, but the primary risk of not talking about and acting on ethics issues is that bad stuff can happen and can continue to happen in the future. What kind of bad stuff? Well, use your imagination! Here are a few themes to get you started: physical or emotional harm, lost time, lost money, lost services, broken trust. Here's a maybe less obvious risk. Avoiding talking about and taking action to address concerns or issues robs you of the opportunity to grow your skills, and pretty much every behavior analyst we know values growth and continued learning. As for the benefits, well, they are pretty much the opposite of the risks. Regularly talking about and taking action with a focus on ethics also helps us build fluency with ethics topics and related analytic skills for you and, *bonus*, for those you are talking and working with!

Antecedent Strategies—Guess what? You are already well on your way to implementing a host of strong antecedent strategies that facilitate talking about ethics. Yeah! Engaging with the content in the intro sections, particularly in Chapter 3, has already moved you closer to getting familiar with and leaning into talking about ethics. Week 2 and you are already winning! Here are three other strategies: (1) read books about how to have challenging conversations, (2) reflect on challenging conversations you have had in the past and try to replay them from the other person's perspective, and (3) prepare for difficult conversations you are likely to have by drafting out a few scripts to get tricky conversations started. In terms of proactive strategies for taking action on ethics topics and concerns, again, you are ahead of the game. Your commitment to engaging with this content and identifying your values can help you lean in to getting things done. Other strategies include using to-do lists, putting tasks on your calendar, breaking tasks into smaller ones, and soliciting support from others.

Red Flags—The red flags that you might be at risk of not talking about and not acting on ethics-related issues are all about your learning history, preferences, and current environment. You've already done some great self-reflection on many of these things. Here are some red flags that come to mind: being conflict avoidant, feeling like you aren't smart enough, being disorganized, disliking ethics or challenging topics or tasks, and being in a toxic or unsafe environment. The idea is to take some time to identify red flags specific to you and your environment that could be barriers to regularly and openly talking about ethics and/or carrying out related tasks.

Addressing—If you know that you struggle with ethics in general, challenging conversation, or task completion, then you are in luck! Being a behavior analyst means you have a wide range of technologies at your disposal to help you address any areas you have identified. You might also consider enlisting the support of a mentor, supervisor, or colleague. The buddy system is a great way to work on some of these skills!

Weekly Actions

MONDAY—THINK ABOUT IT

- When was the last time you talked about ethics with someone in the workplace? Okay, no fair because we sure hope that you did that last week! Before you picked up this book, when was the last time you talked about ethics? Who was it with? Why?
- Take a few minutes to think about your behavior related to talking about ethics. Do you find this topic difficult to talk about, or do you get amped to chat about ethics? How about your behavior relative to challenging topics? Do you try to avoid or soften difficult conversations, or do you come in hot, with a sometimes overly direct approach? None of these approaches are wrong, and knowing your general approach can help you be proactive in making slight adjustments for increased success.
- Think about the models you have had in your personal and professional life. Maybe you have seen some excellent models of how to initiate and navigate tricky topics in conversation. It could be that you had professors and supervisors who talked about ethics in a measured and exciting way. On the other hand, you have just as many examples of individuals dancing around important topics, being very defensive, or taking an overly aggressive approach. Maybe your models for talking about ethics were folx who framed ethics as boring, overly rigid, or even unnecessary. Again, it's not that any of these models are better than others—it's more important how you choose to behave in relation to them. So, taking a few moments to really think about your past experiences can help you identify some of your areas of strength and for improvement.
- Think about how you incorporate ethics into your supervision of supervisees and trainees. How often does it come up? Do you address it just when ethical dilemmas arise, only as one section of your supervision to meet requirements, or throughout the supervisory relationship?
- Notice how you feel and what thoughts come up for you when you spend some time thinking about talking about ethics in general and about ethical dilemmas in particular.

- Jot down notes and thoughts:

TUESDAY—LOOK FOR IT

- Take the day to look for where and why ethics is talked about in your workplace. Maybe this happens regularly because your organization has a structured team or a specific role that focuses on ethics. If so, cool! If so, are there ways that you think that team or person might facilitate even more ongoing discussions about ethics? Maybe ethics is only talked about when an issue arises. Maybe ethics isn't talked about openly at all. Don't judge; just notice.
- If conversations happen about ethics in the workplace, how do they feel? Are they stressful or calm? Do only certain people or people in certain roles have the conversations, or are they open and accessible to everyone?
- Do you see ethics-related tasks occurring in your workplace, and if so, what types? For example, when processes are developed or revised, is there a proactive consideration of ethics content? Do supervisors regularly check in with their direct reports to determine if there are any emerging or present ethics concerns?
- In your workplace, is there a formal process for bringing up ethics concerns, nominating improvements, or getting support to address an ethical dilemma? When team members do bring these things up, what's the response? Is the response dismissive, supportive, or punishing?
- Jot down examples:

WEDNESDAY—TALK ABOUT IT

- Start a conversation with your mentor or supervisor about ways to talk about ethics and difficult conversations that are inviting and reinforcing to the communication partner. Consider asking them if they ever struggle with how to talk about ethics content and initiate tricky conversations.

- Get a convo going with your supervisor at work about the things you noticed from the prompts on Tuesday. If you feel comfortable, share concerns and ideas with them and ask how you might be able to be involved with any tasks (provided you have the capacity and interest). If you did identify areas for improvement, do your best to explain your concerns clearly AND offer some possible solutions.
- With your supervisees and trainees, start talking about talking about ethics and difficult conversations. Use some of the prompts from Monday and Tuesday to get started and to challenge them to reflect on.
- Jot down notes and thoughts from your conversations:

THURSDAY—ACT ON IT

- You might have ended last week without a to-do list, or maybe there were a few things on it. Not to sound like jerks, but we hope that you are sliding into the end of Week 2 with a developing to-do list. Take some time to work on your to-do list. Indicate the things you can do this week and maybe by yourself, and then identify the things that might take time, might need help from others, or might need to be passed along to leaders.
- Jot down thoughts and tasks:

FRIDAY—REFLECT ON IT

- Reflect on how you were feeling at the end of Week 1 as compared to how you feel now. Have your thoughts and feelings shifted? Maybe your resolve to engage in this practice has just strengthened.
- From this vantage point, it may feel like making it through the remaining 50 weeks is going to be a slog. Maybe it even feels impossible. That is totally understandable. Believe us, sometimes writing the content felt that way! Now is a good time to reflect on your values and connect with the ways you outlined for leaning in and behaving in alignment with them.

- Now, we want you to reflect forward. Having gone through these last 2 weeks, what barriers do you anticipate coming up that will make it difficult to stay the course and engage in your daily ethics practice? We don't want you to imagine doomsday scenarios. We do want you to anticipate external and internal factors and contingencies that might be road bumps. Why? Well, what do you do when you see a road bump coming up ahead of you when you are driving? You slow down. You might even angle your tires to minimize being jostled around.

  Same with your daily ethics practice. Maybe you know you are moving soon, changing positions, helping care for a loved one, or taking a much-needed vacation. So, reflect on the barriers that you think life will likely throw at you, and plan for how you will respond. Maybe you'll pause your practice or slow it down. Maybe you'll commit to sticking with it and slowing down or pausing something else. The idea is just to be proactive.
- Jot down something new you learned or something that surprised you:

Resources to Check Out

- Brodhead, Quigley, and Cox (2018) wrote a lovely article outlining strategies for assessing an organization's ethical values. They wrote the article for behavior analysts interviewing for positions, but we think that their guidance can be helpful as we work to engage with noticing our own behavior and the behavior of those in the workplace.
- There are three fantastic articles that you might consider reviewing that all focus on developing a structured approach to building systems and processes aimed at addressing ethics content in the workplace from Brodhead and Higbee (2012), Cox (2020), and LeBlanc, Onofrio, et al. (2020).

Wrap up—As you roll to end of your second week, try not to get overwhelmed. Breathe. This practice requires you to keep your eyes on the horizon. We will likely repeat this several times, but think of this as a marathon, not a sprint. That is precisely why we took the time to walk you through what to expect and how to prep yourself, so that you are well fueled and ready to begin or ramp up your daily ethics practice. When you come back to this book on Monday, you will get started on your first biweekly pair focusing on ethics related to scope of competence. Engage in some self-care this weekend, and we will see you on Monday!

Weeks 3 & 4—Knowing & Evaluating Your Scope of Competence

Week 3

Check in—Hey, hey, hey! This is your first biweekly pair that focuses on an ethics topic. Hopefully you feel prepared and sort of know what to expect after reading Weeks 1 & 2. We hope that you have arrived at this week feeling excited. You might also be a bit nervous, and that is perfectly okay. This first topic is probably pretty familiar to you and is one of the more straight-forward ethics topics.

Focus for the Week—Your scope of competence is all about the boundaries of your professional knowledge and skills as a behavior analyst at a given point in time. Scope of competence applies to the professional activities that you engage in—things like therapeutic services, training and supervisory activities, and the development of systems or processes. The critical thing is that you've got to *know* what those boundaries are for you.

To "know" your scope of competence, you've got to describe and evaluate the knowledge and skills that you currently have and use clearly, honestly, and specifically. The next step is to review data and ask: Do you consistently and accurately use your knowledge and skills to produce the desired outcomes? You might be thinking, *Great! But exactly what am I evaluating*? Well, things like your general clinical skills (e.g., developing common acquisition or reduction programs, implementing common assessments); specific clinical skills (e.g., developing unique or uncommon acquisition or reduction programs, implementing specialized assessments); and the application of those skills to populations, settings, and areas of need. Figuring out your scope of competence is not a one-time evaluation. Just as you change and evolve over time, so does your scope of competence.

Values Check—You (hopefully) read the section on values in Chapter 3. Think back to the values you identified to serve as your compass on your journey. Which one is most related to this topic? Maybe it's the value of growth, learning, improvement, or integrity. As you move through the activities, anytime you feel yourself pulling away from the material or maybe thinking, *Nope, that's not me. I've got this on lock*, we invite you lean in and focus on your related value or values as a way to keep engaging with the content.

Risks & Benefits—There are very real and serious short-term risks associated with not being able to accurately evaluate and understand your scope of competence. If you don't know your current scope of competence, you could implement interventions, agree to provide supervision to a trainee or supervisee in a practice area you haven't mastered, agree to provide services to a client or in a context you don't have solid skills in, and someone could get hurt. For example, if you haven't mastered the skills necessary to work with individuals who engage in severe self-injurious behavior, that client could experience physical and emotional harm because of your actions. Insufficient client progress could impact access to funding for services and the caregivers' trust in ABA services.

Some of the benefits of evaluating and knowing your current scope of competence include your clients making desired progress because you have a well-matched caseload; enjoying your work and keeping stressors manageable; keeping clients, supervisees, and trainees safe; and identifying areas where you can grow professionally. Those are all great things, right?

Antecedent Strategies—The best antecedent strategy for knowing your scope of competence is to reflect on it and then describe it. That's really it; write it down! Given that your scope of competence will naturally change over the course of your career, a description at one point in time isn't enough—you need to regularly evaluate your skills and knowledge. A helpful antecedent strategy is to schedule regular check-ins with yourself and your supervisors or mentors to make sure that your description is still accurate. Put recurring events in your work calendar for a quarterly check-in. During these check-ins, review data (e.g., client, caregiver, and supervisee/trainee outcomes and satisfaction) and evaluate if you're consistently achieving the desired outcomes.

Review procedural integrity data that others have collected for you, or at least that you can collect for yourself, to check how accurately you've been implementing things like assessments, teaching strategies, caregiver and team training, and other important administrative or case management tasks (e.g., reports, records). Always consider the match between your scope of competence and the needs of a client, caregiver, supervisee, or trainee anytime there is going to be an addition to your caseload.

You can also add stimuli into the workplace to prompt everyone to think about their scope of competence. For example, include training about evaluating their scope of competence in onboarding activities for trainees accruing fieldwork experience hours. Team members can be prompted to provide a description of their perceived scope of competence to be reviewed in regular team member performance evaluations. You could also gently probe an interviewee's understanding of their current scope of competence by asking questions like "How might you describe the scope of clients you feel competent to provide services to?" and "What is one thing you would like to add to your scope of competence?"

Red Flags—Two broad indicators that there might be a risk are (1) how confident you are relative to your actual competence and (2) how accurate of a self-observer and evaluator you are. Thinking about these things can make us feel uncomfortable, and admitting that you see one or both of these red flags isn't easy. If you lean on the side of being overconfident or severely under-confident or are inaccurate at identifying the level of your skills, then you might be at a higher risk of struggling to accurately evaluate or describe your current scope. If you have data that suggest your efforts are not consistently producing the desired outcomes, you might not have an accurate understanding of your scope of competence. If you regularly engage in thoughts like "I really should not be doing this" or "I have no idea what I am doing," you might have a case of imposter syndrome. But you might also be tacting that your skills do not match the needs of the situation. These are all red flags that you should watch for in your supervisees and trainees and teach them to keep an eye out for themselves.

Addressing—If you have identified that you do not have a clear understanding of your current scope of competence, or that you have difficulty evaluating it, there are some specific actions that might help. You might spend some time reading articles or book chapters covering scope of competence, self-observation, and self-evaluation. Check out the resources at the end of this week.

Just sitting down and listing your work experience and reflecting on the consistency of positive outcomes related to variables like populations and characteristics (e.g., age, diagnosis, presenting needs, identity), settings (e.g., in-home, clinic, school, community), assessments and interventions, and training and supervision can be very helpful.

It can also be useful to compare your self-evaluation with an evaluation completed by a supervisor or mentor and then discuss the points of match and mismatch. Of course, these strategies are also good for teaching your supervisees or trainees how to address issues related to scope of competence.

What if you think a colleague or supervisor is struggling with accurately knowing or evaluating their scope of competence? Well, that calls for an open and compassionate conversation that you should be well prepared for after completing this week's activities.

Weekly Actions

MONDAY—THINK ABOUT IT

- How do your values align or show up related to knowing and evaluating your scope of competence, and how can you use your values to stay motivated to make reflecting on it a regular part of your professional practice?
- Think about your learning history related to knowing and evaluating your scope of competence. Was this directly or loosely modeled or taught to you?
- Can you identify risks or harms that occurred because you (or others) did not know or continually evaluate your (or their) scope of competence in the past?
- If someone asked you right now to briefly describe your current scope of competence and how you evaluate your scope, could you do so without hesitation?

- Jot down notes and thoughts:

TUESDAY—LOOK FOR IT

- Can you identify any examples of how knowing your scope of competence comes up in the workplace for you and others? Are there antecedent strategies in place for evaluating, describing, and considering scope of competence? Are there any risks or harms occurring related to you (or others) not knowing your scope of competence?
- Can you identify examples of where it could be beneficial to add in content related to knowing and evaluating your scope of competence in the workplace (e.g., hiring, client case assignment)?
- Jot down examples:

WEDNESDAY—TALK ABOUT IT

- Talk to a colleague, your supervisor, or your mentor about this content. Consider asking something like: "I've been reflecting on scope of competence and specifically about how important it is to know and evaluate our current scope. Do you ever think about that?" Ask them to share how they evaluate and describe their scope of competence. If they don't have an answer at the ready, ask if they would be willing to try writing a description of their scope of competence and sharing it with you later in the week. Offer to share yours (you'll work on that tomorrow). Discuss the variables you used in your evaluations.
- Talk to your supervisees or trainees about this topic. You can use the content from the week to have this conversation. If they're ready, ask them to make a list of the things they feel competent in and then discuss it with them in the coming weeks, focusing on matches and mismatches based on your evaluation.
- Jot down notes and thoughts from your conversations:

THURSDAY—ACT ON IT

- Make a list of the variables you should consider when evaluating your scope in terms of practice areas and outcomes. Use those variables to complete a self-evaluation of your scope of competence.
- Based on your self-evaluation, write a clear description of your current scope of competence. Ask your supervisor, mentor, or trusted colleague to review your description and explore areas of agreement and disagreement.
- Make a to-do list of the relevant antecedent strategies to review, revise, or create as needed. For example, maybe you can add (or ask to add) collecting a statement of scope of competence to the application, performance review, or case assignment processes.
- If you identified any red flags with yourself, or your colleagues, supervisees, or trainees, make a plan (with due dates) for how to address them using some of the strategies provided or that you came up with.
- Jot down thoughts and tasks:

FRIDAY—REFLECT ON IT

- Reflect on the things you identified in your workplace and the conversations you had. As the week closes, do you feel that you have a better understanding of how important it is for behavior analysts to know and evaluate their current scope of competence?
- Think back to Monday's question that asked if you could easily describe your scope of competence, and compare it to the description you wrote yesterday. How has your understanding of your current scope of competence changed over the week?
- What will you do to stay active with knowing and evaluating your scope of competence? What will you do to actively teach others to know and evaluate theirs?
- Jot down something new you learned or something that surprised you:

Resources to Check Out

- Brodhead, Quigley, and Wilczynski (2018) have a lovely article that discusses scope of practice versus scope of competence and includes strategies for determining your own scope of competence. Be sure to check out the table with questions to help in your self-evaluation.
- APBA provides scope of competence resources to members, including a document that discusses what scope of competence is and provides practical strategies for when and how to conduct a self-evaluation, tips for how to support supervisees and trainees in learning how to self-evaluate, and a set of companion worksheets to guide you through taking a structured approach to your continual self-evaluation and planning (APBA, n.d.).
- Spencer and colleagues (n.d.) wrote about scope of practice and competence in the context of interprofessional collaboration between behavior analysts and speech-language pathologists. They include some helpful graphics to illustrate these concepts and where overlap sometimes occurs between these two professions.

Week 4

Check in—Alright, you've had a full week of thinking and talking about scope of competence and planning out some to-dos. Take some time to check in with yourself and gauge how you're feeling about this topic. As you reflect on this topic, you might find examples where you think you may have practiced outside of your scope. You might realize that you haven't been regularly evaluating your scope of competence or that you haven't been expressly teaching trainees and supervisees how to do it. That's okay. Be kind to yourself, and recognize that this is an opportunity to learn and grow. That's what this week is all about! In addition to actively integrating ethics content related to scope of competence into your practice, don't forget to read through the scenario.

Scenario—Read the scenario and complete the prompts. Remember, the goal is not to identify the specific standard from an ethics code that was violated, although that is a great activity! The purpose is to think critically about *any* situation and see the ethics-related content and actions.

Carlos, a behavior technician working under the supervision of Yvette, was on vacation for 2 weeks. During those 2 weeks, Yvette added a new skill acquisition program for a client using a teaching strategy that she did not have in any other client's programming. During that time, she also asked the technicians present at the clinical meeting to rate their scope of competence with the teaching strategy, assessed their skills, and provided training to mastery to the two techs who did not demonstrate initial competency. When Carlos returned to work, they read through the program list and the notes from the clinical meeting they missed and initialed at the top, as required. Since they had never heard of the teaching strategy in the new program, they did not run that program and figured that Yvette would pop in to train them. Carlos kept an eye out, but no Yvette. After the session, Carlos left a note for Yvette saying, "Hey Yvette—so good to be back! Everything in the session went well. That new program looks really cool, but I did not run it, as it is not within my scope of competence. When can we meet so that I can get trained on that program?

- **VALUES CHECK**—If you had to take a guess about Yvette's values as a supervisor, what might one or two of them be, and how can you tell based on the scenario? Do they align with your values?
- **IDENTIFY RISKS**—Circle language indicating a risk that was present, and then write down the risks/resulting harms.

  Did Yvette or Carlos identify the same risks? Did they take action to minimize or avoid the risk? Are there other risks or actions you can think of that could be related to this situation?

> ***Be sure to consider specific aspects of the scope of risks or harms!***
> - ***How big/bad is the risk/harm?***
> - ***Who is/would be impacted?***
> - ***Was the risk/harm possible, or did it actually occur?***
> - ***Is it likely that the actual or possible risk/harm will occur again in the future?***

- **FIND ANTECEDENT STRATEGIES**—Go back through the scenario and draw an arrow to indicate anything you think functioned as an antecedent strategy. Make a list of the antecedent strategies, and evaluate if they were effective at preventing things like risks or actual harms. How do you know? Could anything have been improved or added?

- **IDENTIFY RED FLAGS**—Underline any red flags that you find, even if Yvette or Carlos didn't notice them. If the person didn't notice them, write down what factors may have contributed to their ethical blindness (e.g., time constraints, lack of knowledge, authority, common practice, perceiving that there was no risk, lack of oversight). Can you think of any others?
- **FIND INSTANCES OF SOMEONE ADDRESSING THE ISSUE**—Put an * next to any language indicating that Yvette or Carlos took specific action to address the issue. Make a list of those actions and then describe if they were effective or not, how you know, and anything you would have changed or added. Were there other points in the scenario where you think one of the individuals could have taken a specific action to address the situation? If so, what would that look like, and how would you know if it was successful?
- **CONNECT**—Reread this scenario and ask yourself, "What would I need to change to make this a clear ethics violation?" Then ask yourself, "Is there anything that Yvette should do differently next time to improve things?" In other words, what would this scenario have looked like if Yvette actively attended to opportunities to behave in a way that maximized protecting clients and team members? Have you ever been in a similar situation or known someone who was? What was the outcome? What would you do differently, or recommend the other person do differently, now?

Taking Care of Business—Take a little time early this week to map out any tasks you identified. These could be tasks from Thursday or new tasks that have come up for you based on your reflections. For example, maybe you don't engage in regularly scheduled self-evaluations, so you could put quarterly reminders on your calendar for a year. Perhaps you decided you want to make a list of when it's critical to consider your scope of competence and then add to it regularly. Or maybe you realized that you haven't been systematically teaching trainees and supervisees to evaluate their scope of competence. Maybe your tasks will be to search for some resources and draft an outline of how you will address this skill and evaluate the outcomes. Some tasks might be completed quickly, like getting reminders on your calendar or adding an item to a standing agenda. Other tasks might require that you break them into subtasks that you complete over several months. That's okay! Remember, the idea is to engage in small, regular actions that bring the topic of scope of competence into focus for you regularly, even daily.

Words of Inspiration—Taking a good, hard look at your skills, focusing on identifying your strengths and areas for improvement, can be a little scary. But you've got this! Making space for honest self-evaluation will allow you to take a thoughtful approach to your professional development.

Weeks 5 & 6—Building Collaborative Foundations With Caregivers

Week 5

Check in—Hey there! You spent time last week in some heavy self-reflection related to your scope of competence as a behavior analyst and reflecting on some action items. Great work! You may find that you reflect on and assess your scope of competence as you interact with content in the biweekly pairs that feels less familiar or topics that were not sufficiently included in your coursework or supervised fieldwork. Looking in the mirror with an honest lens can be one of the most challenging activities that we engage in, but also one of the most rewarding. And, spoiler alert, we will ask you to do this a lot throughout this book. We are going to further expand on reflection activities this week to focus on the rapport we develop with caregivers. Here we go!

Focus for the Week—In our work as behavior analysts, we typically provide services to an individual, but that individual has an extended network of others who influence their environment in a way that can support or detract from the services we provide. We have an ethical responsibility to facilitate effective coordination of care with this extended community, in partnership with the individual we're serving. The better we do this, the better we can expect the outcomes to be. This week, we'll explore how to establish the initial relationship with this extended network, which may include but is not limited to the initial welcome, intake, assessment, and treatment plan review. In Weeks 21 & 22, we'll dive into how to manage the ongoing partnership with an ethical lens.

Let's start by engaging in some perspective-taking for a moment. You're a parent or caregiver. You have a 3-year-old child. For the last year and a half, you've gone through questions in your head, with your friends, and with your child's physician about what is "normal." You received clarity 2 weeks ago: Your child received a diagnosis of autism. Now you find yourself sitting in front of someone called a behavior analyst, not really sure what that means or why you're sitting in an office across from one. You feel anxious and uncertain. You're processing what was and what might be, but you're here and have a sense of hope for what can be. This is a critical moment in the life of this caregiver, and you, as the behavior analyst seated across the table, have the opportunity to make it great.

There is a quote by Henry David Thoreau that highlights the many arts and beauties in this world, but the highest and most precious art is the ability to "affect the quality of the day" (Thoreau, 2016, p. 97). You're likely working through this book because you're not only a behavior analyst but a service provider as well. In your role, you can positively affect the quality of the day for those you serve. What an incredible privilege that is. We have an obligation to include families and caregivers in treatment from the beginning and throughout the duration of service delivery. But is going through the motion of calling to schedule a caregiver training sufficient? Does doing a cursory review of a behavior change procedure before implementation check off that box?

There's a difference between going through the motions to say we did it or document that we tried, and showing up in a way that caregivers truly feel heard and valued. We have a responsibility to not just do, but to behave in a way that ignites and inspires collaboration and optimizes the benefits for their child. You've likely heard the old adage, "It's not what you say, it's how you say it." Well, it's also true that it's not just what we do, but how we do it. How we behave is a language all its own: the language of kindness, the language of compassion. When we communicate effectively, it serves as a vehicle to move faster, more effectively, and more efficiently in partnership with caregivers. We challenge you to dive deeper than the service-level actions of going through the motions and into how you actively and collaboratively partner with families to fully support their engagement and access to care.

Values Check—As we discussed earlier, there's a good chance this area aligns with your values to help others, and to do no harm. Take a moment to connect with your values related to this topic and reflect on your own history of behavior with effective care coordination. There are a few questions that may be helpful to ask, as you increase your awareness or define your values in this area.

- Do you tend to check off boxes and go through the motions when you initially meet a family and client, or do you tend to lean in and listen to their experience, ask for their input, and integrate that feedback to craft your recommendations and next steps?

- Do you believe that a strong therapeutic relationship creates a foundation and infrastructure to build from and support better outcomes?
- What are your thoughts around family input and your role in treatment? Is the goal to collaborate and compromise within assessment and treatment, or to dictate the recommendations and plan to the family?

Reflect on the thoughts that you had as you read these questions. You may have strong and well-defined values already, or this may be an opportunity for value creation to integrate into your practice. Your responses may suggest that your values are aligned with independence, accuracy, or perhaps prioritizing your degree of expertise and experience. You may give more weight to your knowledge and experience as compared to what the caregiver brings to the table, including their experience, knowledge, and opinions. Perhaps your answers highlight values that do both: You recognize this as a two-way partnership that's critical to any successful outcome.

Remember, your values are your own, and they may change over time. As you work through this section, challenge yourself to not only define your values but also evaluate the extent to which you live them. Are your current behaviors reflective of what you indicate your values are? Do you engage in changes to get one step closer to aligning your values and the actions that you take?

Risks & Benefits—Okay, we're going to do a little more perspective-taking to set the stage, so get comfy! ☕ We typically seek medical care because we need the support of a professional to help us identify a problem and find a solution. We may experience feelings of fear, vulnerability, and confusion because a part of our health and well-being is threatened by something that we can't control. We need help and we seek help from a provider who can resolve our concerns and support our path forward.

Think about a time when you needed medical care and had an exceptionally positive experience with the provider. Since we value different qualities and things, the reason that your experience was exceptional will vary from someone else's. But there are likely consistent themes in how the provider listened, attended to your needs, asked questions, took time to be thorough, addressed your concerns and feedback, explained recommendations and the plan, and integrated your feedback in a way that provided reassurance that they care. If this was your experience, you likely contacted reinforcement and engaged in behaviors to access the care that you needed.

The importance of an effective therapeutic relationship from the onset of services is that it lays a strong foundation and sets the expectations for what is to come. The risk of not doing so is the missed opportunity to establish an effective therapeutic relationship and integrate families as active participants in programming and treatment. The greatest risk to a poor caregiver relationship is they might not access treatment at all.

The beginning of services is a critical time in the lifespan of services because we establish the rules of the partnership and map out the blueprint or plan. When a strong relationship is not established, we risk creating a plan that is substandard to what could be, because it misses key components to success. This may include failure to obtain comprehensive and accurate information during the initial intake appointment, ineffective communication because the caregiver does not trust the clinician or the creation of a plan that does not include all the critical pieces of information to appropriately address needs. Penney and colleagues (2023) state that behavior change without input from caregivers is not, by definition, ABA, and there is nobody better to identify the areas of concern and need for clients than their families.

The benefits of beginning services in partnership with caregivers are endless, and they build and compound over time. We are more likely to get accurate information, identify goals that are meaningful for the client and their context, and support the caregivers and other natural change agents in learning strategies that are practical for their lives. This partnership can be a source of reinforcement for caregivers and others and increase the likelihood of access to care and the related lifelong benefits.

Antecedent Strategies—Our goal is to establish a strong and effective relationship with caregivers from the beginning of services, to ensure that we have the right information to create the best plan and recommendations for the individual in our care. We can use antecedent strategies to minimize any risk that may threaten the foundation of a growing and collaborative therapeutic relationship. We may be the experts of behavior analysis, but the family is the expert of their child. We should consider the following strategies to create a space and partnership that integrates the value that each caregiver brings.

1. **Establish expectations early:** At the onset of services, caregivers and providers should outline, discuss, and document what is expected of the other. This can happen vocally and in writing and should happen multiple times during the relationship, not just at the start. This may include initial paperwork including a consent for services, information about the program and services, a policy that outlines expectations of family engagement and participation in services and the outcome of not doing so, scripts for team members to utilize outlining who to reach out to on their care team for different needs, how to escalate concerns, and the frequency of communication from both the provider and family. Openly and proactively sharing expectations cultivates an environment of trust and collaboration that is critical to the onset of this partnership.
2. **Ask the right questions, and don't forget to listen:** Our ultimate goal is to help those we serve. A good place to start during the initial assessment process is to utilize resources to comprehensively evaluate the individual's needs. Asking questions and listening to understand are critical skills that can help build trust and ensure that we have the information that we need to inform the most appropriate recommendations and plan. The plan should be a combination of our knowledge and expertise, as well as the expertise of the family, including the targeted goals and programs, intervention, and ongoing progress monitoring. If we attempt to create a plan in isolation, in the absence of this partnership, we've missed the opportunity to fulfill our ethical responsibility to involve caregivers.

3. **Collaborate and communicate the plan:** Have you ever given a friend feedback about not doing something and then realized that you never actually told them the details of what they should have done? Oops! We've all been there, but we should strive to avoid this with families. Clear communication and thoughtful, collaborative planning are the keys to avoiding breakdowns and misunderstandings. We want to set up every caregiver to be wildly successful, and communicating the plan in a clear, consumable, and inviting way is critical to that. We have a responsibility to provide information in a way that is heard and understood, so that caregivers can make informed choices (Penney et al., 2023). Schedule a time with each caregiver before initiating any behavior change protocol—not just because it's your ethical responsibility, but because you won't be nearly as effective if you try to implement it alone. It means that you only have half the team showing up to the game, and to win, you need the whole team there, knowledgeable of the plan, excited, and prepared to win.
4. **Have the right skills, space, and time:** You don't typically go to a job interview so confident that you have the right skills that you don't prepare at all. So, give that same importance and time to your new clients. You should ensure that you're just as prepared when you show up to complete an initial assessment for a family in services. Based on preliminary knowledge, can you effectively meet their needs? Do you have the skill set and knowledge to effectively support them? Have you selected an environment to complete the initial assessment that's welcoming and supportive of what needs to be accomplished? Have you communicated where to be, what to expect, and how long it will take? Are you prepared to allocate your mental effort and energy fully to the family you're meeting with to ensure that all key information is incorporated into the assessment and treatment plan recommendations?
5. **Engage in compassionate care:** Remember, it's not just what we do, but how we do it. As we integrate caregivers into treatment, we must explain behavioral strategies and recommendations in a way that is understandable and that shares the power of decision-making. If something doesn't feel right to a caregiver, don't

launch into convincing them. Instead, make space for learning why. Excellent customer service, kindness, and care serve as a gateway to establish access to ethical practices. Put another way, it's nearly impossible to involve caregivers in treatment if they don't access reinforcement through an effective, strong, healthy therapeutic rapport with their care provider. There are excellent resources that we encourage you to check out that highlight the importance of using soft skills and how they're often the most important skills.

6. **First impressions matter:** A caregiver's first impression with behavior analysis impacts not only that experience or organization, but the field. And most importantly, the opportunity for their child to access quality care and lifelong benefits. The benefits of a good first impression are lasting. The established rapport serves as the foundation for engagement, partnership, and success. Behavior change requires data collection, treatment integrity, and consistency, which can prove very challenging for families with multiple children, jobs, financial burdens, and other common life factors. We want to minimize the risk that comes with a lack of engagement and increase the likelihood of a strong partnership by *showing* the benefits, not just talking about them.

There are specific expectations that come with "firsts," so we should attend to the opportunities and make them great. The first caregiver meeting after starting services is a tremendous opportunity to influence a positive experience and maintain the momentum of engagement that has been established.

- *Welcome them with a smile, and greet them all by using their name.*
- *Be an active listener as demonstrated in your demeanor, body language, receptiveness, and follow-up questions.*
- *Hold yourself accountable to listen to understand and not simply respond during the conversation. Pausing before you interject can be a helpful strategy to ensure you're holding yourself accountable.*

- *Throughout the meeting, take turns asking, sharing, discussing, and celebrating. As active participants, we should ask:*
    - *How things are going*
    - *What they are most excited about*
    - *What they are fearful or worried about*
    - *How their child is doing from their perspective*
    - *What questions they have about the specific treatment and protocol*
    - *What targets would make sense to incorporate based on natural contact in their home environment*
    - *How we can make the ongoing meeting most valuable for them*
    - *When they are interested in participating during sessions*
    - *And the list goes on!*

The key is that we meet caregivers where they are and we walk with them; we support their understanding and engagement in a way that develops their repertoire to support their child. We are building up a team that supports an expanded infrastructure for the individual we serve, and the stronger the framework, the better the outcomes. Remember, this is not one-sided. A partnership goes two ways. How will you pull from your individual pools of knowledge and bring them together to a common understanding that supports where they are and where they can go?

Be sure to show more than you tell. You can tell families that services are important, that services should be prioritized above everything else, and assign trainings and tasks. You can also make them a part of their child's first word, drinking from a cup for the first time, or walking independently through a store for the first time with no concerns or challenges. If there is minimal engagement, ask yourself when the family experienced the last benefits of services. Lastly, never end the current interaction without having a plan for the next. You have a great family meeting, but when is the next scheduled? Is it recurring, so that everyone can plan accordingly and look forward to the agenda that brings mutual benefit?

Red Flags—While antecedent strategies proactively help us to stay on course, red flags are a quick sign to indicate that we are off course and at risk of not fulfilling our ethical responsibility. It can be difficult to lean into hard things, but time is of the essence, and the identification and resolution of red flags is critical, especially when it comes to effectively managing therapeutic rapport. There are several red flags to have on your radar, but the first is probably if your values do not include collaboration or a similar value. Another is if your training and experience have not included an emphasis on compassionate care and effective therapeutic rapport. We encourage you to connect with and use the resources shared in this section and reach out to a mentor who can provide you with feedback and additional guidance.

A specific red flag found in practice is if you find yourself saying things like, "I just sent them the initial paperwork and consent for services to sign; we don't need to review it." This indicates a missed opportunity to establish expectations and understanding from the very beginning, and there will likely be miscommunication in your future. Additional red flags may include caregiver meetings that have no structure, agenda, or focus, or meetings that are overly structured and allow no space to ask open-ended questions and hear feedback and concerns from the family. Hearing a caregiver share "I have no idea what my child is working on" is a huge red flag suggesting a big miss with our responsibility to include caregivers in practice and communicate programs and progress.

Anything surprising (and not in a good way) can be red flags, such as, "I had no idea that I was expected to consistently attend scheduled sessions, participate in services, and attend caregiver meetings. I have to work, and this is better than daycare." If you work within an organization and have limited access to resources and minimal-to-no training on how to establish effective relationships and the policies and procedures do not align, those are also red flags. It's difficult to drive home the importance of caregiver engagement if you are not in an environment that values and fosters this. If you have a supervisor, does your supervisor ask about caregiver engagement and satisfaction? If not, that is likely a red flag. Is there sufficient training in place to conduct initial assessments, clinical documentation, treatment plans, and an audit to review the integrity of those completed activities? If not—you guessed it—that's another red flag.

Addressing—Take some time to review your notes and self-evaluation and then make a list of what you will act on. What's in your control to change? Perhaps you'll work on your own active listening skills and review the articles listed on compassionate care to put them into action. You may not have a new individual starting services, but you can reflect on the experience of your last client intake to inform your next. Did the interactions, communication, and action align with your ethical responsibility to involve caregivers from the onset of services in a way that established a strong foundation to build from? If not, what will you do differently next time?

Before the next time, perhaps you'll improve fluency with the clinical intake documents to ensure that you review the content in detail and answer any questions the family has, prior to signing. You may expand the assessment resources that you use to include a caregiver stress survey and ensure their concerns and priorities are captured in the programming and interventions. Maybe you recognize opportunity in how you explain the importance of attendance and ongoing participation with families, because you reflected on several red flags that have risen with other families. So, you plan to proactively incorporate this discussion the next time you welcome a new family to care.

If you work within an organization or with others, there is likely an opportunity to review how the processes, policies, and resources make it easier to do the right thing and minimize any vulnerability, to most effectively engage in ethical practices related to caregiver engagement. You could review the quality of the intake documentation, who completes it, and what training they receive to see if there are areas for possible improvement. Maybe you'll work with your organization to ensure the right assessments, interviews, checklists, and scripts are in place to ensure the relationship starts strong and a comprehensive review is completed to ensure the right treatment is in place. If all this great structure is in place, there may be the opportunity to request additional training, modeling, and mentorship to support the treatment plan review and initial communication with families.

Weekly Actions

MONDAY—THINK ABOUT IT

- Based on your reflection and analysis of how well you and/or your organization prioritize an effective caregiver relationship when getting started with services, you may have a few things to do. It's easy to point fingers at others, so let's start with you.
- Reflect on your current relationships and rapport with caregivers. Many of them may be well-established, but how do you approach getting started with new families? Do you take a structured approach, or do you wing it?
- Think about a new individual you are or will be serving and if you have the knowledge, resources, and support to establish a strong therapeutic relationship from the onset of care.
- The best way to communicate the benefit of ABA services to families is to "show" them, rather than "tell." We want them to live and experience the benefits, so how do you engage and structure the environment to ensure this happens?
- Reread the "Values Check" section above, and think about how your values show up in your relationships with caregivers. Do you show up as the person and professional who you want to be and live the values that you have identified?
- How do you impact the quality of the day for others?
- Jot down notes and thoughts:

TUESDAY—LOOK FOR IT

- Observe those around you, and look for examples and non-examples. Who appears to engage effortlessly with families and have strong caregiver participation?
- Where do you see the missed opportunity? There is missed opportunity everywhere because we have every opportunity to inform the experience for others. Think back to the perspective-taking activities and how you want to experience care as a patient. Where do these expectations show up in your behavior as a service provider?

- Where do you see this done very well? When others are welcomed, valued, and heard? How do you know?
- Jot down examples:

WEDNESDAY—TALK ABOUT IT

- Talk to a colleague, mentor, or supervisor about your reflections. Try something like, "I know the importance of caregiver engagement, but my behavior and priorities did not always reflect this. How do you think we can better structure our resources, meetings, and communication to make it easier to prioritize caregiver engagement at the onset of services?"
- Speak with your supervisor about what you think is going very well and what opportunities exist with both your cases and the organization's structure and resources.
- Meet with your expanded clinical team and/or supervisees to engage in perspective-taking based on the current structure of your process and expectations.
- Spend some time talking to your supervisees and trainees about the importance of building strong collaborative relationships with caregivers from day one. Consider reviewing the risks, benefits, antecedent strategies, and red flags from the sections above and facilitating a discussion.
- Jot down notes and thoughts from your conversations:

THURSDAY—ACT ON IT

- Once you discuss your action items with relevant supervisors and individuals in your organization (if you work for one), implement some great changes! Perhaps you talk with families about their experience and introduce surveys at the beginning of services, and maybe again 30 days later.
- Address the red flags you have identified, even if you just tackle one from your list. What accountability have you put in place to follow through and ensure that you are supported in that process?

- Try to observe others reviewing initial intake paperwork, the initial family interview, and the assessment. Consider reviewing each other's treatment plans and completing a 360° review.
- Show up today differently than you did yesterday, and behave in a way that aligns with your values.
- Jot down thoughts and tasks:

FRIDAY—REFLECT ON IT

- Over the course of this week, you have used a different lens to view interactions with caregivers, the importance of including them as partners in service delivery, and how your behavior influences and informs the extent to which that can happen. What did you experience and take away from this increased awareness? How is your perspective different now from what it was last week? What will you prioritize differently, and how will you translate these learnings into long-term behavior change for yourself? How will you continually evaluate the effectiveness of your approach and modify it to make it even better?
- We talked a lot about red flags, but what are some green flags that you see that indicate you are doing this exceptionally well? You may hear things like praise and feedback from parents about their child's progress, requests for more meetings to learn more strategies, and them sharing success stories with their friends and referring others. You might see families consistently attending sessions, caregivers actively participating in sessions and across multiple settings, quick responsiveness to emails and questions, and active participation in the creation of the initial treatment plan and priorities.
- Jot down something new you learned or something that surprised you:

Resources to Check Out

- Check out *How to Win Friends and Influence People* by Dale Carnegie (2022). He outlines specific recommendations to help develop and maintain positive, effective relationships.
- In "Translating the Covenant," Foxx (1996) provides perspective and a reminder to behavior analysts to be aware of the language we use to ensure that it can be heard and understood by the listeners. Read this article and think about how you talk to caregivers and other collaborators. Do you expect them to become fluent with behavior analysis jargon? Or can you fluently explain things using everyday terminology?
- LeBlanc, Taylor, and Marchese (2020) do a beautiful job highlighting the importance of compassion in our work as behavior analysts and how it translates into action as we cultivate and support compassionate and effective relationships with caregivers.
- Chapter 9 in the book by LeBlanc, Sellers, and Ala'i (2020) focuses on strategies for effectively building therapeutic relationships and is chock full of resources to support you in your efforts.
- Check out the gem of an article by Wolf (1978). It's a classic that tells us to reflect on the heart of behavior analysis, which is to help others in ways that are meaningful to them.
- Penney et al. (2023) encourage us to consider compassion as an eighth dimension of behavior analysis, which we review in greater detail during Weeks 21 & 22.
- Grenny and colleagues' (2022) *Crucial Conversations* is a great resource to help you prepare to identify when conversations become critical and how to navigate them.
- Kirby and colleagues (2022) discuss the need for behavior analysts to engage in "humble behaviorism," and while it is focused on interprofessional relationships, the concepts can be applied to relationships with caregivers.
- Taylor and colleagues (2019) outline skills to teach soft skills for engaging in compassionate care with caregivers. They also identify barriers that may impact a behavior analyst's ability to build a successful relationship with a caregiver.

- Rohrer et al. (2021) created the *Compassionate Collaboration Tool* to support behavior analysts in providing compassionate care in their service delivery.

Week 6

Check in—Awareness and self-reflection can be a superpower. The more we do it, the better we get. You spent the last week reflecting on your values in the context of caregiver partnerships and the importance of effective therapeutic rapport. Your behavior and caregiver relationships may be a gold standard that others can look to as an example. Or perhaps you're feeling a bit overwhelmed with your learnings and opportunity. The good news is that you are more aware today about what you don't know, and you have identified tasks to get you closer to where you want to be. Either way, you should be proud! You are engaging and showing up in a way that is better than yesterday, to better those around you.

Scenario—As you read the following scenario, we don't want you to just categorize what is "ethical" or "unethical," but instead reflect on the extent to which the individuals' behavior influenced an effective relationship, in a way that minimized risk to both the family and professional. Let's take a behavior-analytic approach to take a look at what was done really well and what potential opportunities exist.

> *Daniel is a certified behavior analyst who is supporting an initial assessment for a new client, Arya. Daniel has been in contact with Arya's father, Noah. Arya is a 3-year-old girl who was recently diagnosed with autism. Daniel calls Noah the day before the in-person assessment to ensure that he has the time and address of the center. When Noah and Arya arrive at the center, Daniel comes to the lobby to greet them. Daniel looks at Noah in the eyes with a big smile and shakes his hand and then gets down on the floor to Arya's level to greet her as well, but with a big high five.*
>
> *They spend about 10 minutes conducting a tour of the center and familiarizing Arya with some new toys. She is very interested in a baby doll, which she takes back to the conference room where they're meeting. Daniel explains to dad Noah that they will spend*

time going through initial documentation and an interview before spending more time with Arya. A second behavior analyst introduces herself to Arya and Noah, and she and Arya proceed to the playroom while Noah and Daniel meet.

Daniel begins the conversation by reiterating how very excited he is to meet Arya and the importance of a two-way partnership to support the best possible outcomes. He indicates that he may be the ABA professional, but Noah is the expert of his daughter, and they will be most successful with his input and support. He asks Noah if he has any questions before getting started, to which dad answers yes. They walk through a series of questions and reservations that Daniel compassionately listens to while maintaining eye contact, nodding his head, asking follow-up questions, and validating Noah's feedback.

Daniel takes time to review the initial clinical documentation, the parent handbook, and the importance of family participation in every aspect of service delivery. He explains how the initial assessment is their starting point and creates a road map that they can use to check in and ensure that they are staying on track to their ultimate goal of discharging from services. He also discusses the importance of data and how those points are used as landmarks on the map to ensure that they are on track and can course correct as they need to. Noah indicates that he did not realize the extent of involvement that is required, but after speaking with Daniel, he is excited. He is concerned that Arya's mom, Rebecca, has concerns with ABA services. Daniel disregarded the comment and continued the assessment with Noah and Arya. At the end of their meeting, Daniel indicates that he will schedule a follow-up time to finalize their assessment before coordinating the next steps.

- **VALUES CHECK**—How did you feel reading this scenario? Does Daniel's behavior suggest that his values align with the importance of effective therapeutic rapport? What did Daniel do that is in alignment with your values?

- **IMAGINE RISKS**—Circle any areas where you could imagine the events playing out differently and inviting risks.

Be sure to consider specific aspects of the scope of risks or harms!
- *How big/bad is the risk/harm?*
- *Who is/would be impacted?*
- *Was the risk/harm possible, or did it actually occur?*
- *Is it likely that the actual or possible risk/harm will occur again in the future?*

  Think about what Daniel did to minimize the likelihood of risk. What did he review, how did he engage, and how did he behave in a way that ensured a strong therapeutic rapport from the beginning to inform the initial assessment, treatment plan, and recommendations? Imagine the risk that would be presented if he did not have access to the resources that he reviewed with the family. What risk would be present if he did not engage in the way that he did?
- **IMAGINE ANTECEDENT STRATEGIES**—What antecedent strategies did Daniel put in place? How did Daniel show up with this family in terms of his verbal behavior and communication? Was the language that he presented consumable to ensure that Noah could access and understand the information? What other strategies could have been used? Draw an arrow to the antecedent strategies that were used, or where there were missed opportunities.
- **IDENTIFY RED FLAGS**—Underline any red flags that you see, even if Daniel didn't notice them. If you found a red flag (which you should have!), what other potential red flags should Daniel have on his radar?
- **IMAGINE INSTANCES OF SOMEONE ADDRESSING THE ISSUE**—Not all scenarios will include a description of someone addressing the concern or issue. Put an * next to a place where you think Daniel could have taken better action. Should he have addressed the reported concerns from the mother differently?
- **CONNECT**—Have you ever been in a similar situation or known someone who was? What was the outcome? What would you do differently, or recommend the other person do differently, now?

Taking Care of Business—Write down your thoughts and frame where you are and where you want to be and go with no judgment. Show up better every day, and start today. What is your plan going to be? What got you excited that you may act on first? Pick something easy to implement that will move you forward to further evolve and develop your initial partnerships with caregivers. Then, keep going.

Words of Inspiration—As service providers, the way we show up and impact the experience for those we serve is an art. It's beautiful and humbling to impact the life of another human for good. Use your canvas well and create a masterpiece that reflects grace, kindness, and positive impact.

Weeks 7 & 8—Knowing Your Supervisory Responsibilities for a Strong Start

Week 7

Check in—Well, you have completed two biweekly pairs, on scope of competence and collaborating with caregivers from day one. How are you feeling? Are you still excited? Less nervous? A little bored? In whatever way you show up this week, we are so glad you are here!

Focus for the Week—There are a lot of pieces that go into providing effective supervision. You've had your own experiences receiving supervision during your training and education that have shaped your supervisor knowledge, skills, and practice. You probably function as a supervisor in some capacity in your role now, and we bet that you've spent time reflecting on the supervision you provide and the relationships with your supervisees and trainees.

And if you haven't reflected yet, guess what's coming up this week? As we think about the supervision we've received and we provide to others, we should be thinking about things like how supervision was and is designed and provided, how feedback was and is provided and received, and how cultural responsiveness played and currently plays a role in supervision.

But wait, there's more! There are a few things that might not immediately make their way onto your list of critical considerations but that are vital to your supervision. Things like staying up to date with supervision requirements, actively managing your supervision workload, clearly articulating the conditions of supervision, and maintaining documentation systems might feel like things you do "automatically." And let's be honest, they probably feel tedious and boring. But they really are important, so let's talk about why.

You likely need to meet certain requirements to provide supervision (e.g., organization, licensure, or certification requirements). You've got to know, stay up to date with, and meet those (sometimes changing) requirements. It's also important to make sure your supervisees and trainees are aware of the requirements. If you fail to meet and maintain the requirements, there could be serious consequences for you and your supervisees and trainees (more on that later).

Before accepting a new supervisee or trainee, it's important to take a moment and assess your capacity to add to your caseload. You might be thinking *But we're talking about adding a supervisee or trainee, not a client!* True. But just pause for a second and think about everything you need to do when you have a supervisee or trainee, and a *new* supervisee or trainee at that. No, seriously. Take a few minutes to stop, think, and make a list.

Did you make your list? It might look similar to the steps you go through when taking on a new client, including activities like spending time building rapport and getting to know them, assessing their current skill level, and discussing and identifying goals, to name a few. There's a potential risk of harm if you commit to taking on a new supervisee or trainee but then find that you've overcommitted yourself and cannot carry out those critical steps, like providing the support and training they need.

It's also vital to communicate the conditions for the supervisory relationship. Outlining clear expectations also allows you to discuss why the relationship might end and what will happen if that's the case. We'll go into that in more detail in Weeks 48 & 49, but you need to communicate this at the beginning of your supervisory relationship. Your supervisee or trainee will be practicing under your direction, and you both must understand the logistics of supervision and your shared ethical obligations. Remember, as their supervisor, you are responsible for their professional work. That means that it's possible that if they do something that harms a client, you may be held accountable as their supervisor if you did not provide appropriate training, supervision, and support.

Finally, although it's not glamorous, creating, maintaining, and storing documentation of your supervision activities is a must. And your documentation system must follow any relevant regulations or requirements (e.g., organization, licensure board, certification body). You wouldn't provide services to a client without a clear service agreement and regular documentation of data related to the assessments and interventions you carry out. So, same thing with supervision (we'll cover more about documentation in Weeks 32 & 33).

Together, these activities all amount to you taking accountability for your supervision.

Values Check—This topic may feel pretty straightforward, and you may feel like you won't need to link it to your values. But sometimes it is the simpler or discrete topics that trip us up, as we are at risk of setting ourselves on autopilot or cutting corners. So, take a breath and spend a few minutes connecting this topic to your values. What value or values come up when you think about making sure you're meeting supervision requirements? What comes up when you think about making sure you have the capacity to take on a new supervisee? What about making sure that you clearly communicate your expectations with your supervisees and trainees? Is there one value that is the best fit for this ethics topic? Keep that value (or values) in your mind as you read on, and ask yourself *How can I use this value to help me identify behavior that is not in alignment with it and behavior that is well aligned to this value?*

Risks & Benefits—Failing to take accountability for your supervision can have both long- and short-term risks. If you don't meet the requirements of a certification body or licensure board, it may mean that your supervisee or trainee loses some or all their experience hours, which could delay or prevent them from moving forward with pursuing certification or licensure. This can cost them money, time, and trust, and can negatively impact their career. Mistakes meeting supervisory requirements for technicians can also result in incorrect implementation of programming or temporary disruptions to service delivery, negatively impacting clients.

If you take on a supervisee or trainee but don't have the capacity, you're likely to provide inadequate supervision. Low-quality supervision can negatively impact the development of their clinical and supervisory skills, exposing current and future clients and others to the risk of harm.

Unclear communication of the supervision conditions might result in miscommunication, frustration, and distrust. Failure to clearly outline the supervision conditions can really complicate things if your supervisee or trainee is not meeting expectations and you have to terminate the relationship.

Some of the benefits of staying accountable for your supervision include having the capacity to provide high-quality supervision, your supervisee or trainee meeting requirements to become licensed or certified, and setting up your supervisee or trainee to be more likely to provide high-quality services and supervision in the future. Across the board, you'll also reap the benefits of developing a trusting and collaborative relationship where you both know what to expect and your supervisee or trainee feels supported.

Antecedent Strategies—

There are a few things you can do to set yourself up for success:

1. *Know the requirements and set up a system for checking for updates. Here's a quick test: Right now, can you list out the initial and maintenance requirements to provide supervision? Don't be too hard on yourself if you can't, but definitely take some time to get fluent with them. Going forward, consider scheduling 15 minutes each quarter to make sure that you are up to date with your knowledge and any required forms. Make sure that update emails from certification and licensure bodies are getting to you (i.e., make sure you keep your email address updated in your profile) and that they're not going into your junk folder, never to see the light of day.*
2. *Assess your capacity to take on a new supervisee or trainee before starting each supervisory relationship. This involves reflecting on your current commitments and evaluating them against the time and effort required to provide supervision*

for an additional supervisee or trainee. In addition to assessing your work requirements, you should also check in on your work-life balance and identify if you have any personal factors that might make increasing your supervisory duties difficult. It's probably a good idea to consider doing this on an ongoing basis (e.g., quarterly) and before taking on additional clients. You might want to look at Weeks 3 & 4 on evaluating your scope of competence as you assess your capacity. We'll also look at assessing your personal biases and factors in Weeks 15 & 16, which will also be valuable to consider when assessing your capacity.

3. *Review the requirements with your supervisee or trainee and have a plan for reviewing any updates with them. If you are providing supervision under a contract, you can review requirements at the same time. For technicians or others you are supervising, it's still important to take the time to do this. And don't assume that technicians or others who have been practicing for a while know these requirements or are up to date on the latest. So, if you're taking over for another supervisor, it's a good idea to just check in with supervisees to make sure they are familiar with their responsibilities.*
4. *Have a plan for discussing expectations with your supervisee before starting supervision. This may be done when you complete a supervision contract or just as a conversation during your initial meeting. And guess what? If you're already providing supervision but never talked about expectations, you can still do it! In fact, just blame it on us and do it in your next supervision meeting.*
5. *Have a plan for how you will create, maintain, and store any required supervision documentation. Early in the supervision relationship is a great time to discuss this with your supervisee or trainee and agree on how they will maintain their documentation as well. It's also a good idea to have a plan to engage in regular audits of your documentation and your supervisee's or trainee's documentation to make sure it's accurate and com-*

plete. Doing so allows you to catch and address errors early, before they grow into big problems.

Red Flags—One indicator that you might be at risk of slipping on your supervisory accountability is if you find yourself thinking *I don't have time to prepare to take on another supervisee or trainee.* This would be a good time to honestly assess your caseload volume (including your supervisory caseload) and determine if you're able to add another supervisee or trainee. Other red flags include not having documentation systems, thinking that it is your supervisee's or trainee's responsibility to take care of their own documentation, and never reviewing supervision requirements. Finally, if you didn't receive high-quality supervision, that is a big ol' red flag that your supervisory repertoire may need a little work.

Addressing—If you have identified that you don't have a plan for addressing these items when providing supervision, there are specific steps you can take to get started. You might take some time to find all the requirements and review your current supervision systems. You might also consider reaching out to a colleague or mentor to chat about what systems they have in place to support their supervision practices.

Taking some time to list out your caseload obligations (e.g., programming, observation, supervision meetings, administrative tasks, treatment plan updating) is also a great place to start. It might also be helpful to have a supervisor or mentor review your list with you. You might also consider practicing with them how you communicate and document the supervision conditions and expectations so that these conversations are fluid.

You might have instances when you or your supervisee need to report deviations in supervision (e.g., you were unable to meet requirements due to unforeseen circumstances) to the relevant bodies (e.g., organization, licensure board, certification body). Or you might have instances when you identify issues with your trainee's supervision hours that may result in those hours not being counted. You might be feeling a little icky thinking about these scenarios. You might also be thinking that you would want to avoid these conversations at all costs. Both of these scenarios will likely result in a difficult conversation to discuss the next steps (e.g., reporting, updating hours) and how to avoid this in the future. But remember, when you do this, you're setting a great example for your supervisee or trainee on how to address these scenarios in the future and ensuring that their supervision meets the relevant requirements.

Weekly Actions

MONDAY—THINK ABOUT IT

- Take a moment to reflect on your supervision experience as a trainee or supervisee. Were these things modeled for you? Were you explicitly taught how to take these steps to support your supervision practices? Or have you been learning by trial and error now that you are supervising? If you find that your self-reflection results in identifying areas where you can improve in the future, it might feel uncomfortable, but know that this is a good thing! You are honestly reflecting on your experience and behavior, and that is going to allow you to improve your processes for the future.
- How do your values support you in your commitment to ensuring that you have the capacity to take on a new supervisee or trainee and that you can meet the relevant requirements (e.g., from licensure boards, certification bodies)?
- What is your learning history related to evaluating your supervisory capacity, keeping up to date with requirements, communicating supervisory conditions, and documenting your supervision? Have you taken on a supervisee or trainee and then found that you don't have the time or resources to support them appropriately?

- Can you identify risks or harms that occurred because you were not prepared to take on another supervisee or trainee? For not communicating the supervision conditions and expectations clearly? For not documenting your supervision activities?
- How do you review the supervision requirements and expectations with your supervisees and trainees?
- Do you know where to access the relevant information about supervision requirements? When you get emails and newsletters from relevant organizations (e.g., certification body, licensure board), do you make time to carefully review them for changes or updates to supervision requirements?
- Do you regularly audit your documentation system and the documentation system of your supervisees and trainees?
- Jot down notes and thoughts:

TUESDAY—LOOK FOR IT

- Can you identify any examples of how accountability in supervision comes up in the workplace for you and others? Are there antecedent strategies in place for preparing you to meet supervision requirements, evaluate your caseload volume, communicate conditions and expectations with your supervisees and trainees, and document your supervision?
- Can you identify examples of where it could be helpful to add resources and support to ensure you're accountable for the supervision you're providing?
- How are trainees and technicians added to caseloads in your organization? Does your supervisor check in with you about your workload?
- Are any data collected and reviewed related to accountability and documentation in supervision? For example, does your organization collect data on the number of supervision deviations for technicians each week, month, or quarter? If so, how are those data used?
- Jot down examples:

WEDNESDAY—TALK ABOUT IT

- Talk to a colleague, your supervisor, or your mentor about this content. Consider asking something like: "I've been thinking about accountability in my supervision practices, how I can make sure I am meeting all the requirements, and how to evaluate my caseload to make sure I can take on a new supervisee or trainee. How do you approach those things?" Ask them how they ensure they evaluate their caseload to make sure they can commit to taking on a new supervisee or trainee. Also, consider asking how they prepare to start supervising a new person and how they document their supervision. You might even ask if they can share samples of how they document their supervision.
- Talk to your supervisor about your current processes for accountability, and ask for their ideas for improvements. You might also consider asking how they approach the conversation if they identify that someone is not able to meet the supervision requirements or is not accountable in their supervision practices.
- Talk to your supervisees and trainees about supervision requirements and their familiarity with these topics. This also might be a good time to make sure they are familiar with the expectations in your supervisory relationship. You might even consider recruiting some feedback on how you shared this information with them (or didn't) at the start of the supervisory relationship. And if the feedback is that you didn't, now is a good time to make sure you do!
- Jot down notes and thoughts from your conversations:

THURSDAY—ACT ON IT

- Take some time to review your process for making sure you and your supervisees and trainees are up to date on the relevant supervision requirements. If you don't have one, take some time to create one. Plan to spend time checking over your process for reviewing expectations with your supervisees and trainees. Ask a colleague, mentor, or supervisor to review the process with you. Then, make sure to review anything you may have missed initially with your supervisees and trainees (e.g., updates to requirements, expectations).

- Make a list of caseload obligations that you should consider when evaluating your caseload capacity. Did you make a list as you were reviewing the "Focus for the Week" section? Make sure to include that in your list. Remember, your caseload not only includes the number of clients you have, but also supervisees, trainees, and administrative tasks that you've got to complete on a daily, weekly, monthly, quarterly, and yearly basis. Complete a self-evaluation using this list. Based on your evaluation, identify what your current caseload volume is. Then, ask a colleague, mentor, or supervisor to review your evaluation and discuss any areas of agreement or disagreement.
- Review your processes for creating, maintaining, and storing your supervision documentation. This is a great time to compare it to the requirements of the relevant bodies (e.g., organization, licensure board, certification body) to make sure you're meeting them.
- Schedule time with your supervisees and trainees to review their processes for creating, maintaining, and storing their supervision documentation. Use this time to make sure the relevant requirements are met.
- Make a to-do list of antecedent strategies related to accountability in your supervision to review, revise, or create as needed. For example, maybe you can discuss with your supervisor that you will evaluate your current caseload as a quarterly goal and a task to be completed each time you consider adding a client, supervisee, or trainee to your caseload.
- If you identify any red flags in your review, make a plan (with due dates) for how to address them using the strategies provided or that you identified.
- Take a breath! We mean it! We've given you a lot of prompts for possible tasks, and we know how busy you are. Don't forget that you'll have next week to work on tasks too. And you can always stay with this topic for as long as you need before moving on to the next biweekly pair.

- Jot down thoughts and tasks:

FRIDAY—REFLECT ON IT

- Take a moment to reflect on the things you identified and discussed this week. As the week comes to a close, do you feel that you have a better understanding of how to be accountable in your supervision practices?
- Think back to the Monday reflection. How has your understanding of your accountability in supervision changed over the week?
- What will you do to stay active in maintaining your accountability in supervision and ensuring you have capacity to take on additional supervisees and trainees if the opportunity arises?
- How will you keep up with supervision requirements as they change? How will you monitor your caseload and other clinical obligations as they change over time? How will you ensure that you're communicating the conditions and expectations of the supervisory relationship? How will you audit your and your supervisees' and trainees' supervision documentation?
- Jot down something new you learned or something that surprised you:

Resources to Check Out

- The book on supervision and mentorship by LeBlanc, Sellers, and Ala'i (2020) covers how to set your supervision up for success in Chapter 2.
- Garza et al. (2018) provide tools for setting up successful supervision.
- The article by Sellers, Alai-Rosales, and MacDonald (2016) discusses the importance of meeting the ethical requirements in supervision, including the responsibility of supervisors around supervisory volume and communication about supervision expectations.

- The article by Sellers, Valentino, and LeBlanc (2016) also provides recommendations on how to create a successful supervisor-supervisee relationship.
- BACB Supervision Resources
    - The *RBT Handbook* (BACB, 2022c) outlines supervision requirements, including how it should be structured and the documentation requirements.
    - The *BCaBA Handbook* (BACB, 2022a) outlines the requirements for supervised fieldwork to apply for certification and the ongoing supervision requirements.
    - The *BCBA Handbook* (BACB, 2022b) outlines the requirements for supervised fieldwork to apply for certification.
    - The *Maintaining Your RBT Certification* (BACB, 2023a) video covers the supervision requirements for maintaining the RBT credential.
    - The *Supervision Checklist for RBTs* (BACB, 2022d) provides reminders for RBTs of tasks and actions to take before, during, and at the end of the supervisory relationship.
    - *Supervision Checklist for RBT Supervisors and RBT Requirements Coordinators* (BACB, 2022e) provides reminders for supervisors of RBTs of tasks and actions to take before, during, and at the end of the supervisory relationship.
    - The "Better Understanding Supervision as an RBT" episode (Jenkins & Ulrich, 2023) of the *Inside the BACB* podcast discusses the ins and outs of RBT supervision.
    - The "Taking Supervision Seriously" section in the August 2020 *BACB Newsletter* (BACB, 2020b) discusses supervision issues supervisors and supervisees might commonly contact.
    - The *Consulting Supervisor Requirements for New BCBAs Supervising Fieldwork* document (BACB, 2018a) provides the requirements and the required documentation.
    - The *2022 Supervised Fieldwork Summary* video (BACB, 2021a) provides recommendations for documenting fieldwork hours.
    - The *Fieldwork Checklist and Tip Sheet* (BACB, 2019) provides reminders for trainees of tasks and actions to take before, during, and at the end of their fieldwork hours.

  - Check out the BACB website for more resources, including a sample supervision contract.

Week 8

Check in—You've now spent a week thinking and talking about accountability in supervision and how to keep up to date with supervision requirements, evaluate your capacity to supervise, communicate conditions and expectations in supervision, and document your supervision. Go ahead and pause and take a moment to check in and see how you are feeling about this topic. You might identify that there have been times when you have been supervising and might not have been meeting requirements or didn't really have the capacity to provide high-quality supervision. And you know what? That's okay. Take a moment and acknowledge that you're not going to be perfect. But what is important is that you lean into your supervisory accountability and make space to continue improving your systems. And listen, making mistakes is a gift that gives you a chance to model for your supervisees and trainees that it's okay not to get it perfect every time, how to own our mistakes, how to learn from them, and the steps to take to make sure that we do better in the future. Now let's look at a scenario.

Scenario—Read the scenario and complete the prompts. Remember, the goal is not to identify the specific standard from an ethics code that was violated, although that is a great activity! The purpose is to think critically about *any* situation and see the ethics-related content and actions

> *Natalia has just been asked by her supervisor, Idris, to provide supervision for a trainee who has just started a master's program and needs to start accruing supervision hours. Natalia is already supervising three trainees, and she has a caseload of 12 clients. Idris said that the trainee should make her job easier, since they will be able to help with writing programming and training the technicians for her clients.*
>
> *After reviewing her weekly calendar, Natalia decided to make a list of all her weekly tasks, including the amount of time needed for each task. She completed this for a full month and identified that if*

she took on another trainee, she'd have minimal flexibility to assist with any client issues, and it would impact the amount of time she had allotted for completing administrative tasks like documenting supervision, updating treatment plans, and completing other projects around the center. Natalia documented all her concerns and prepared to discuss them with Idris in their next meeting.

In the conversation, Idris thanked her for documenting all of her tasks and assessing her capacity. He asked Natalia to review her caseload and identify what could be removed or changed in her caseload in order to allow her to take on another trainee and said they would follow up in their supervision meeting the next week.

Idris shared that as it was a requirement of Natalia's position to provide supervision, she may need to find a way to fit it into her caseload. Idris reminded Natalia that it would benefit the organization long term if they could support trainees with accruing experience hours and that the organization had a goal of having a cohort of at least 10 trainees each year.

- **VALUES CHECK**—If you had to take a guess about Natalia's values as a supervisor, what might one or two of them be, and how can you tell based on the scenario? Do they align with your values? What about Idris's and the organization's values?
- **IDENTIFY RISKS**—Circle language indicating a risk was present, and then write down the risk/resulting harms.

  Did Natalia identify potential risks? Did she take action to minimize or avoid the risk? Are there other risks or actions you can think of that could be related to this situation? Did Idris identify potential risks?

Be sure to consider specific aspects of the scope of risks or harms!

- ***How big/bad is the risk/harm?***
- ***Who is/would be impacted?***
- ***Was the risk/harm possible, or did it actually occur?***
- ***Is it likely that the actual or possible risk/harm will occur again in the future?***

- **FIND ANTECEDENT STRATEGIES**—Go back through the scenario and draw an arrow to indicate anything you think functioned as an antecedent strategy. Make a list of the antecedent strategies, and evaluate if they were effective at preventing things like risks or actual harms. How do you know? Could anything have been improved or added?
- **IDENTIFY RED FLAGS**—Underline any red flags that you find, even if Natalia did not notice them. If she did not notice them, write down what factors may have contributed to her ethical blindness (e.g., time constraints, lack of knowledge, authority, common practice, perceiving that there was no risk, lack of oversight). Can you think of any others?
- **FIND INSTANCES OF SOMEONE ADDRESSING THE ISSUE**—Put an * next to any language indicating that Natalia took specific action to address the issue. Make a list of those actions, and then describe if each was effective or not, how you know, and anything you would have changed or added. Were there other points in the scenario where you think Natalia could have taken a specific action to address the situation? If so, what would that look like, and how would you know if it was successful? How do you think the follow-up meeting with Idris might go? How should Natalia address the issue if Idris continues to push that she has the capacity to provide supervision to the trainee?
- **CONNECT**—Imagine how the follow-up meeting with Natalia and Idris might have gone and ask yourself *How would Idris respond that might make this a clear ethics violation? How might Natalia respond in this situation? How would Idris respond that would not be a potential ethics violation?* Imagine this situation did not go well on Natalia's end (maybe Natalia didn't document her tasks and assess her capacity), and ask yourself *What would I need to change to make this a clear ethics violation?* Have you ever been in a similar situation or known someone who was? What was the outcome? What would you do differently, or recommend the other person do differently, now?

Taking Care of Business—Take some time at the start of this week to create a plan for completing the tasks you identified. They might be tasks from your reflection time or tasks you identified on Thursday. For example, maybe you don't have a consistent method for documenting your supervision sessions, so you need to schedule some time to evaluate what you're currently doing across your supervisees and trainees and create a standardized process. Or maybe you recognized that you haven't been discussing supervision requirements with your supervisees and trainees at the start of supervision and reviewing them on a regular basis to make sure you're meeting them. You could add it to the meeting agenda once a quarter with your supervisees and trainees. And maybe you're also scheduling time on a regular basis to review the requirements on your own to make sure there haven't been any changes or updates.

Some of these tasks might be quick to complete, while others might take some time and reevaluation to make sure what you've put in place is working. Remember, the goal is to make sure you're accountable in your supervisory practices and continue to review your practices regularly so that you don't inadvertently cause harm to supervisees, trainees, or clients.

Words of Inspiration—When you acknowledge your mistakes and show how you grow from them, you make a brave space where supervisees and trainees can learn and grow as well. That is the essence of supervision—continual evolution to become our best selves in the service of others. That is the perfection in the imperfection.

Weeks 9 & 10—Resting & Reconnecting

Week 9—Time to Rest & Take Care of Yourself

This is the first time you have reached a rest and self-care week, so here is a special message for you: **You are doing an incredible job!** We hope that you are engaging in daily self-care and wellness strategies aimed at keeping yourself balanced and healthy. We also acknowledge that we are asking you to do hard work, to think about tough topics, to engage in challenging and often uncomfortable self-reflection and conversations, and to take action.

So, after 8 weeks (or maybe more, if you needed to pause and come back to the material) of hard work practicing how to think, talk, and take action on practical ethics topics, you've earned yourself a scheduled break. Your assignment is to NOT think about ethics. Honestly, if we are doing our job, this will be difficult, or even impossible for you. The whole ding-dang point of this book is to help you build your fluency with thinking about ethics, seeing a bit of ethics in everything you do. So, don't try too hard to not think about ethics; just take the week to move through your days noticing whatever you notice.

Here are some prompts or ideas to help you make the most of this week. You'll run into this rest and self-care week in Weeks 17, 25, 34, and 42, so don't feel pressured to do all these things this week. Just do what feels right; even if that's none of these ideas. And maybe, if you find any of these ideas helpful, you'll consider incorporating them into your regular daily practice.

We've given you a special spot at the end of this section for you to set your intentions, because we really want you to commit to using the week to rest and recharge. Then, the next time you come to a rest week, maybe look back at the past rest weeks to see what you have tried, what's working for you, and what you might want to try. But before you start engaging in purposeful self-care and wellness activities, spend some time reviewing these ideas and checking out the linked resources. You can go back to the Chapter 3 sections on values and on compassion for more ideas and resources. And if you don't use any of the resources, that's fine too. This week is for you, so use it with the intention to create space to rest and recharge.

Ideas for Checking in With Yourself

Behavior analysts and behavior technicians spend most of their time in service of others. We spend time observing others and working to support them. But you need some attention and care too. So, here are some ideas for checking in with yourself.

- ✓ **Check in with your body:** Take a few minutes to do a scan, starting with your head or toes and working in the opposite direction. Notice how each body part feels. Are you holding tension anywhere? If so, try to let go of that tension. Drop your shoulders away from your ears. Unclench your jaw. You can also try contracting your muscles, holding for a three count, and then slowly releasing. Try starting with your toes and systematically moving up and throughout your body to your fingers.
- ✓ **Check in with your breathing:** Notice how you are breathing. When you breathe, is your chest rising and lowering? Maybe your belly is expanding and contracting. Try a few of these breathing strategies:
    - **Diaphragmatic breathing:** Put one hand on your belly and one on your chest. Try breathing in slowly through your nose so that your stomach expands, moving your hand up and down, while keeping the hand on your chest relatively still. When you breathe out, contract your stomach muscles, and breathe out slowly through your nose or mouth while keeping your chest relatively still. Keep practicing until you can breathe into and out of your belly for a slow, steady count of three to four. Try doing this breathing exercise a few times a day.
    - **Box breathing (aka 4x4 breathing):** Slowly breathe in through your nose, expanding your belly and chest, for a count of four. Hold your breath for a count of four. Now slowly and steadily exhale for a four count. Hold for a four count. Repeat these steps three to four times.

- ✓ **Check in with your mind:** Schedule a few minutes in a quiet, comfortable place to just sit and listen to what is going on in your mind. As you sit quietly, are you having recurring thoughts? If so, what's the topic? The idea is just to observe your thoughts, not to judge them, not to challenge or debate them. Just see if you can watch them, the way you might pause and watch the clouds pass overhead. If it helps, you can play soft music or look at an object like a tree. If you find sitting still uncomfortable, you can do this check-in while walking, doing something repetitive (e.g., crocheting, coloring, doodling), or even just standing and swaying back and forth. You might try doing this check-in for just 2–3 minutes the first few times and then increase it as you feel more comfortable.

Accessing Things That Feed Your Soul

Take some time to make a short list of things that you love doing that are calming or healing for you. Maybe it's reading for pleasure, gardening, doing yoga, snuggling your pet, having coffee with a friend, or cooking. Don't make a long list, maybe just three or four things. Remember, you are going to have these rest weeks throughout your practice with this book, so it might be good to rotate across a wide variety of activities and take note of the effects they have on your well-being. Once you have your list, add them to the Monday–Sunday list (keep reading, you'll find it).

- ______________________
- ______________________
- ______________________
- ______________________

Try Something New

Try one new thing this week. Maybe you want to try a new activity like pickleball, painting, or salsa dancing. But you don't have to go big to get the effects of getting outside your comfort zone. Try a new podcast, fruit or vegetable, or restaurant. Heck, you can start small and just try a new dish at your favorite restaurant. Listen to a type of music you haven't listened to before. Take a different route on your walk, bike ride, or commute. The idea is just to shake up your routine, challenge yourself a bit, and maybe notice something new and develop some new hobbies and self-care activities. Spend a few minutes making a list of just a few new things you might consider trying out this rest week and in the rest weeks to come.

-
-
-
-

Look for Awe & Practice Gratitude

What is "awe"? Well, according to scholars, experiences that produce awe have two critical features: (1) We experience the vastness of something else (and in turn perceive our related "smallness"), and (2) we must make a mental shift to accommodate and make space for this new experience (Keltner & Haidt, 2003). Research suggests that experiencing positive awe produces a wide range of beneficial effects, such as lowering indicators of chronic inflammation, increasing our sense of connectedness to others, deepening our humility, and elevating our mood and feelings of happiness (Allen, 2018a).

So, how can you increase the awe you are experiencing? Well, spending time in nature, listening to music, and looking at art can all produce a sense of awe, if you are willing to be present and allow yourself to be impacted. The Greater Good Science Center describes taking a "savoring walk," where you spend 20 minutes walking and actively noticing positive things that you see, hear, or perceive through your other senses (Greater Good Science Center, n.d.-a). When you notice one of these things, pause and really pay attention, becoming aware of the experience you are having.

So, spend a few minutes thinking about things that give you goosebumps and take your breath away, but like, in a good way. Maybe it's holding a baby, listening to opera, or watching the sunset. Maybe it's getting your hands in the dirt in your garden or feeling the wind in your face while riding your mountain bike. Now, use the space below to list some of those things, and see if you can schedule time into your Monday–Sunday list to increase the likelihood that you contact awe throughout your week.

List of your possible awe-inspiring activities/experiences:

- ______________________________
- ______________________________
- ______________________________
- ______________________________

Alright, so finding awe is awesome (c'mon, we *had* to). Turns out that engaging in gratitude also has a bunch of beneficial outcomes! Studies and meta-analyses indicate that the positive effects of adults engaging in gratitude exercises include increased well-being, positive affect, and feelings of happiness and satisfaction (Allen, 2018b). You can even take a gratitude quiz (Greater Good, n.d.).

There are many gratitude practices, including keeping a gratitude journal; writing gratitude letters; engaging in mental subtraction (essentially imagining what life would be like if a specific positive thing in your life did not occur); and doing a specific type of journaling called "Three Good Things," where you write down three things you are grateful for each day and identify the causes of those good things (Allen, 2018b). If you'd like to try journaling, but you are not quite sure how to approach it, Thnx4! is an online journaling platform that even lets you set up or join multiday gratitude challenges (Greater Good Science Center, n.d.-b).

You can access a whole host of awe and gratitude activities, as well as activities focused on practices like compassion, happiness, and mindfulness on the Greater Good in Action webpage (Greater Good Science Center, n.d.-a). Maybe you want to add some time to your Monday–Sunday list to check out their resources.

Words of Inspiration—Now is the time for rest. Rest can be challenging for many of us, so lean into this practice. It is not selfish. In fact, to give yourself time and permission to truly rest is a critical step toward radical self-love. When you take care of yourself, you increase your ability to take care of others.

Monday–Sunday Self-Care & Wellness Intentions

Map out your self-care and wellness intentions for this week and beyond.

____

____

____

____

Get to your comfy spot. Maybe do something special this time, like light a candle, put on your comfiest sweats, or play the most calming or inspiring music you can think of, to really create a space of peace and relaxation. The next step is important—get these things on your calendar, or they might not happen! If you are like us, you might have a little voice in the back of your mind that tells you: *Self-care is selfish and wasteful. You could be getting XYZ done right now.* But self-care is NOT selfish. It is essential, even if it's difficult for you. So, you can tell that little voice:

Thank you, little voice, for keeping me humble. I also know that I am valuable, and I don't have infinite resources. Emily, Sarah, and Tyra said I've got to take time to recharge, so that's what I am going to do!

Think of these intentions as gentle goals to take care of yourself so that you can keep taking care of others. If you can't meet all the goals, that's okay too. We are so proud of the work you have put in so far. We hope you know how amazing you are to make ethics a priority in your already hectic professional life. Now get out of here and go relax!

Monday Self-Care & Wellness Intentions

Tuesday Self-Care & Wellness Intentions

Wednesday Self-Care & Wellness Intentions

Thursday Self-Care & Wellness Intentions

Friday Self-Care & Wellness Intentions

Saturday Self-Care & Wellness Intentions

Sunday Self-Care & Wellness Intentions

Ideas for Regularly Scheduled, Recurring Self-Care & Wellness Practices

Take some time midweek and on Sunday to make note of anything you tried that you really enjoyed, and see if you can find time to regularly integrate that practice into your daily, weekly, or monthly schedule.

Week 10—Time to Reconnect With Your Daily Practice

Hi there! How are you feeling after getting some rest time? We hope that you are returning feeling well rested and recharged. For this week, we just want you to ease back into your daily practice of noticing, reflecting on, talking about, and taking action on ethics content. In the past 6-ish weeks, you have focused on building your ethical know-how related to scope of practice, purposefully building a strong collaborative therapeutic relationship with caregivers, and supervisory requirements.

Start off this week reviewing, reflecting on, and celebrating successes and tasks you have checked off. Take the bulk of the week to refine, plan, and complete existing tasks for these topics. So, the focus of this week is to give you some catch-up time before we move into new territory and add more to your task list. You'll end this reconnect week scanning the weeks ahead so that you can be proactive in your time management and planning of your daily ethics practice.

MONDAY

Review & Reflect— Spend 20–30 minutes flipping back through Weeks 3–8, and review related notes you have been keeping. Think about how you felt when you started on your journey of building your ethics muscles. Think about how you feel now. Have your thoughts and feelings about this content shifted? If so, how?

Take stock of the number of times you reacted to the content with feelings of discomfort, frustration, maybe even guilt or shame. Make note of the number of times you reacted to the content with feelings of confusion, discord, or feeling particularly challenged.

Play back the conversations you have had over the past several weeks. Were you nervous to have them? How did they turn out? Better than you expected? Is it getting easier to think and talk about ethics content? How did it feel to take a week to rest and focus on yourself? Did you identify any wellness and self-care practices to embed in your daily, weekly, or monthly schedule? Maybe jot down your answers to these questions, or other notes from your reflections.

Celebrate— Time to gloat! We want to hear all your accomplishments throughout the past several weeks. Maybe you leaned into a challenging conversation that you might have otherwise avoided. Perhaps you tacted a skill or practice that you are lacking and made moves to build your repertoire. Maybe you improved or created a policy, process, or resource. Maybe you did something nice for yourself. Heck, maybe you just stuck with reading this dang book daily! We so often focus on praising others and beating ourselves up, so take this moment to give yourself some kudos!

TUESDAY THROUGH THURSDAY

Refine, Plan, & Act— How about the tasks that you haven't been able to get to? Looking back, are there some tasks that you planned to do but haven't been able to clear from your to-do list? Jot those down in the space provided below, and spend some time identifying why you haven't been able to complete them. Yeah, we know, "not enough time" is likely the first reason that comes up. Limited time is so real. And there may be other contributors like you might not feel skilled enough, you don't know where to start, or the tasks seem really challenging or feel aversive.

Reflecting on the barriers to completion can help you make a plan for moving forward. Maybe you need time to brainstorm with others, to partner with someone to share the workload, or find a way to make it less blech. Maybe you bit off more than you can chew, and you need to break it into smaller tasks spread over more time. Whatever the reason you didn't get a task done, it's okay! Use this week to make a *reasonable* list of the things that you might be able to knock off your to-do list in the next few days. For the things remaining that you won't be able to get to this week, thoughtfully plan and calendar them across the coming month. Remember, it's a marathon, not a sprint.

Tasks for This Week:

____

____

____

____

____

____

Tasks to Find a Cozy Spot on Your Calendar:

____

____

____

____

____

____

Well, what are you waiting for? Get after it! Use your list to guide you. Use your values to guide you. Use your supervisor, mentor, colleagues, supervisees, and trainees to support you. Using the list that you made above for tasks to complete this week, check them off as you go, and be sure to praise yourself. If you don't get to everything, don't beat yourself up. Take a few minutes to identify why, move them to your future action list, or add them to a spot on your calendar. The key is thoughtful planning to make it happen!

FRIDAY

Look Ahead— Alrighty, you can still use today to wrap up tasks, but we also want you to peep ahead at the coming biweekly pair topics. Maybe they all feel timely, relevant, and doable. Sweet, we are excited for you to connect with the topics and continue to build your daily practice of engaging with ethics content. On the other hand, maybe you have something going on at work or in your personal life that makes some or all of them challenging or just not possible. That's totally fine too.

You can slow down, pause, or skip a biweekly pair that feels too tricky, painful, or just not a good fit for right now. Just commit to restarting or getting back to the topic as soon as you can. Part of our hope for your journey is that you lean into scanning ahead and planning in meaningful ways that allow you to live your values, contact lots of reinforcement, build your skills, take good care of yourself, and do the hard things.

Inspiration to Reconnect— You are doing an incredible job, and the people in your workplace (supervisors, colleagues, supervisees, trainees, clients, caregivers, and other professionals) are so lucky that you are committed to building your practice. You are making our profession stronger, healthier, and safer. And we are so grateful.

Weeks 11 & 12—Protecting Confidentiality

Week 11

Check in—Hey there, you're into the next section already! We hope that you've started, or will soon develop, a rhythm to reading this book in a way that works and makes sense for you and your daily ethics practice. We are glad to jump in with you again. You're doing great!

Focus for the Week—Over the next 2 weeks, we'll discuss a topic that reaches across all that we do—protecting confidentiality. It is difficult to talk about confidentiality without also discussing dignity. Dignity is defined as the state or quality of being worthy of respect. As service providers, supervisors, and researchers, we have a responsibility to protect the dignity of those we serve and can do so in how we manage the information that is shared with us.

There are lots of ways that we treat others with dignity, including how we talk to them and how we implement teaching and intervention procedures. Another way that we treat others in a dignified way is to protect their confidential information by only disclosing what is necessary and remaining in compliance with requests to effectively maintain the information that is entrusted to us. As you move through the content in this section, we encourage you to hold yourself accountable and reflect on the function of your behavior when you share information that does not pertain to you, whether it is related to a supervisee, client, research participant, or family member.

Here are some tough questions that may cause a strong reaction. Ready? What information are you sharing that is not your own, and with whom? Do you share information on a need-to-know basis? Do you share information related to the individual that directly benefits them and facilitates effective care coordination, or is it to gain attention or a reaction from someone? Take a breath because those questions were not easy. Great job honestly answering them so you can increase awareness of your potential areas of risk.

Instead of talking about the definition of confidentiality, let's walk through some synonyms and orient to what we are really talking about: private, personal, classified, restricted, privy, privileged. When you read those words and think about yourself as a patient, student, trainee, or employee and what you would expect from those responsible for you, what are your thoughts? It is important to reflect and ground ourselves in the meaning of confidentiality, to ensure that we never lose sight of the importance and specific impact that our choices can have on the lives of others.

Information is shared with us with the intention to help and make someone's life better. Protecting that information is at the heart of therapeutic and supervisory relationships. We also have an ethical responsibility as behavior analysts, which of course is why we dedicated a whole section to this topic! So, what does protecting confidentiality actually mean? Let's break it down:

1. We have a responsibility to engage in appropriate actions to protect those under our blanket of services, whether it be a family member, client, team member, or research participant. Our actions must include reasonable attempts to prevent the inadvertent sharing of privileged information. That means if we irresponsibly (e.g., without ensuring we have the right permissions, to get access to social reinforcers from friends when we tell a funny story or share about a client on social media) or accidentally (e.g., leave documents open on our laptop while working at a café, attach the wrong treatment plan or supervisee performance evaluation to an email) share confidential information and do not follow the appropriate steps to correct, we are responsible regardless of our intentions.
2. We have a responsibility to know and follow the laws, policies, and expectations related to maintaining confidentiality.
3. We are responsible to maintain confidentiality across all types of information and ways it could be shared including telehealth (most popular in recent years!), video recordings, documentation, data, emails, in person, and phone conversations.
4. We have a responsibility to ensure that we, and our direct reports, take proper care with the information that is entrusted to us when storing, transporting, and getting rid of documentation.

Values Check—How do you think about dignity and identify with it in your practice as a behavior analyst? Dignity is a central part of care services and a long-established practice in the medical field, but how does this resonate with you? When you think about confidentiality, has your history, training, and experience been thorough and focused on the impact to patient care; or, has it been more focused on online modules, trainings, and signing off on policies? It should probably be a healthy combination of both.

Either way, what matters is that you are here and taking the time to know your values and how to prioritize and behave in a way that supports your ethical responsibility! Are your values aligned with doing no harm? How do you perceive confidentiality, dignity, and doing no harm in the context of your role and responsibility as a behavior analyst and care provider?

We want to treat and discuss personal information in a way that respects and never undermines the individuals who we serve. So, check yourself. Are your values aligned? Is your behavior aligned with your values? Do you protect confidentiality with the utmost degree of care in how you communicate, what you communicate, and with whom you communicate?

Risks & Benefits—There are clear risks to violating confidentiality, including legal or other disciplinary actions from certifying and licensing bodies and the organization you work with. Those consequences can include loss of licensure, certification, or employment; and state or federal actions including significant fines. We don't take this topic lightly, and we want to ensure that you're equipped with the knowledge and awareness that you need to make the best possible decisions for yourself and others.

This is also an area that we consider to be a common pitfall, as it is so vast and can become a blind spot if we think confidentiality is no big deal, if we become complacent, or if we think that it is someone else's responsibility. We never want to become too comfortable in our responsibility to protect confidentiality for the family, client, supervisee, or trainee information that we hold. We risk compromising a culture of confidentiality and dignity, which has a damaging impact on the rapport, trust, and future ability to fulfill our responsibility to others. The benefits of fulfilling our ethical responsibility to maintain and manage confidentiality include creating a space of trust, protection, and confidence that care and supervision are protected and respected. This is foundational to patient and supervisee experience and their related outcomes.

Antecedent Strategies—The key to doing the right thing is knowing the right thing that needs to be done. So, know your stuff. Let's first talk about policies and protocols. Are there outlined descriptions of how to manage, maintain, store, share, and get rid of permanent products? Are there similar protocols to describe expectations related to what can be shared, who it can be shared with, and when it can be shared related to all communication modalities? These are critical antecedents to have in place, because without them we are unlikely to do the right thing.

Next up is training! Are team members aware, knowledgeable, and fluent with the expectations that have been outlined? Are they demonstrating competency as evidenced by completed audits? An effective antecedent strategy is to have ongoing formal trainings, in addition to less formal trainings and feedback, to differentially reinforce compliance with confidentiality requirements.

Now, let's think about the space in which you provide services and supervision and have sensitive conversations with and about clients and supervisees. Is the space conducive to protecting the information that is being discussed, or can people overhear you? We recommend taking a fresh look at your home office space (if you have one), where conversations and services occur in a center, and anywhere that you share or access personal information from others, to ensure that you are thoughtful about the space that you are in.

Never underestimate the power of awareness. Know where you are, what you are saying, and why you are saying it. Hold yourself accountable. We know this is a bit direct, but we can never risk becoming too comfortable with information that is not ours. Set rules for yourself, such as "I do not discuss clients outside of the presence of a team member within the same organization"; "I do not take a work call when standing in line at the grocery store, because I could slip up and compromise confidentiality"; or "When I work in public, my screen protector is on and I am connected to private Wi-Fi."

Know your risks and your responsibilities, and set simple rules that make it easier to do the right thing, especially when you are distracted or life is chaotic. It's similar to how you should always put your car keys in your lunchbox so you don't forget it at work, or always park in the same parking area at the airport so you don't forget where you parked. Simple rules minimize our risk for error, and that is very important when it comes to protecting the information shared by others. We also encourage you to hold your colleagues accountable by using checklists or reminders for individuals to lock up information, offering an alternative meeting area, or another great idea that you brainstorm!

Red Flags—Read through these red flags with a lens of what you would hope for and expect from your own service provider or supervisor. You likely go to your provider because you trust and respect them as a professional, their judgment, and their ability to help you. Hopefully it is the same with your supervisor. We think the following red flags are unfortunately fairly common but easy to mitigate. We want to increase the likelihood that the right behaviors will happen, and red flags are typically scenarios in which we don't do a great job of that, and we are inviting some of those risks.

Do you or others use your personal cell phone or laptop? Is your phone and computer protected with a password? Does your entire family and friend group know your password (so it's not actually serving its intended purpose)? Is there a culture of confidentiality in which it is prioritized and at the forefront of daily behavior and workflows, or do people view it as a technicality, or worse, a pain in the rear? Is there a lack of proactive and recurring training around the importance of confidentiality and how to effectively implement precautions? Those are all red flags.

Alright, so we covered a few red flags that could indicate that you or the organization are at risk for future violations. Now let's review some red flag behaviors that are pretty commonplace, are actually violations, and can be indicators of deep problems when they become standard practice.

Have you ever seen a performance review lying on the counter for everyone to see? Or perhaps a corrective action form that outlined the substandard performance, disciplinary action, and feedback for the team member? Have you ever had a supervisor provide you with corrective feedback in the hallway for all to hear? Or observed a conflict in which a supervisor spoke with your peer about your performance? Have you seen other individuals providing telehealth services in public without a privacy screen or in situations in which the client can be seen and heard? How about disclosing a client's medical record to the wrong school or disclosing the record to another provider without a signed consent?

We could write an entire book about the red flags around breach of confidentiality, and much of this is driven by poor awareness, sloppy work, lack of knowledge, poor management, insufficient accountability, and minimal oversight. Even if accidental or the result of carelessness, if these types of behaviors go unchecked, they can bloom into a culture of unethical behavior. These red flags are certainly not comprehensive, but we offer them to make you pause and think. You might see these situations regularly. You might have even done some of these things. And all of that is alright because we are going to walk through strategies that you can incorporate into your daily practice to help improve things. One of the biggest red flags is if there is no clear and appropriate way to escalate these violations of confidentiality within your organization, and/or if there is a culture where people don't feel comfortable bringing up these issues.

Addressing—Alright! So, unless you gave yourself a big green checkmark and pat on the back for doing all the things right, well, you probably have some work to do. That is great! Instead of beating yourself up, we should think about how knowing what is wrong is the first step in the right direction. You know where you are and what you need to do. You can also serve as a model to others around you, in how you show up, take caution, provide and accept feedback, and behave in a way that ensures confidentiality and respects the dignity of the individuals in your care. If you are worried about common practices in your workplace that might violate confidentiality, you can start a conversation with your supervisor. You might say, "I know how busy we all are providing great services. I have noticed that in our rush to get all the things done, we might be slipping on our responsibility to be careful with confidential information. Could we chat through some things I have done and seen and come up with some solutions?"

We encourage you to think through the following weekly actions to own and address next steps. When we think about how, what, and where to prioritize, we should always start with ourselves. So, review and reflect on the greatest areas of opportunity and the resources or knowledge that you need to bridge the gap to compliance. Then, expand to supporting others and engage in bigger and broader change as needed. It is similar to when you are on an airplane, and you are instructed to put on your oxygen mask first before you can help other people!

Weekly Actions

MONDAY—THINK ABOUT IT

- Take some time to review the values section above and your notes, and reflect on where you stand on your opinion to own the responsibility to maintain confidentiality. Then, think about how well you do it.
- How did the connection between confidentiality and dignity resonate with you?
- Has your degree of awareness changed throughout the course of reading this section?
- Has your training included education and an awareness of your ethical responsibility to communicate and maintain these expectations?
- Jot down notes and thoughts:

TUESDAY—LOOK FOR IT

- The red flags that we shared were NOT comprehensive. We've barely scratched the surface. The first thing to really critically think through is all the potential gaps in our behavior to effectively maintain confidentiality.
- Check yourself. Where do you have confidential conversations? How do you transport confidential documentation? How do you dispose of and store confidential information? Do you fully understand what confidential information is?
- Look at the training in your organization. Are new team members trained on the policies and procedures that relate to confi-

dentiality? What is the quality and scope of the training? Are audits carried out to ensure that people are maintaining the things they learned in the training?

- Review the documents in your organization. Are there clear policies and procedures related to maintaining confidentiality and addressing breaches? What do your confidentiality and exchange of information forms look like?
- View the spaces you and others work in through different lens. Are they structured in a way that facilitates confidentiality? In how the rooms are structured, the available spaces, and any prompts or reminders? What about paper and electronic documentation? How is it stored, accessed, and shared?
- Do you see great examples of protecting confidentiality? If you do, give your friend a shout-out and thank you.
- Jot down notes and thoughts:

WEDNESDAY—TALK ABOUT IT

- Talk to a colleague, mentor, or supervisor about your reflections. What do you, your colleagues, and the organization do very well? Where is there opportunity for improvement? Brainstorm together and solicit feedback. Based on your knowledge, what should be done differently, and why?
- Chat with your supervisees and trainees. Do they have a full understanding of confidentiality and their responsibilities? Talk to them about those obligations and link confidentiality to quality treatment, dignity, and healthy therapeutic and supervisory relationships. Maybe you have made a mistake related to confidentiality that you can share with them as a learning opportunity. Ask them if they have seen any issues related to confidentiality and if they have any ideas for improvements.

  Maybe even develop some scripts and role-play using them to help them be prepared for common situations like overhearing a peer talk about confidential information with friends at a party, or how to respond if a neighbor approaches as they are heading into a home session and says, "I see you and others coming and going from that house every day. What are you doing?"

- Jot down notes and thoughts:

THURSDAY—ACT ON IT

- If you have identified some areas in your own behavior, and we are not judging here, make a plan for improvements. For example, if you realized that when you work from a local café, you often sit so that your computer screen is in plain view, or that you are taking confidential calls on speaker in a home office space where others can easily hear the conversation, what will you do to improve your attention to and protection of others' confidentiality?
- Work with your team or friend to create a checklist or to-do list based on what you have found and want to change.
- Talk with your team about what you found, what recommendations you have to do things differently, what ideas they have, and what resources you will need to accomplish your goal.
- Jot down notes and thoughts:

FRIDAY—REFLECT ON IT

- What content did we review this week that hit home the most? How has it changed your perspective, and how will it drive lasting behavior change?
- Are there opportunities for you to share what you learned with others to expand the impact, and to let them share their perspective and impact as well?
- As you reflect on what you learned, challenge yourself one last time if you had thoughts like *I don't agree with that* or *That doesn't apply to me.* Share those reflections with your colleagues and supervisor to keep yourself accountable.
- Jot down something new you learned or something that surprised you:

Resources to Check Out

- Weiss and Russo (2022) have a nice chapter discussing the risks that behavior analysts and technicians face when sharing information on social media. The authors acknowledge the very real challenge of how ever-present social media is in our lives and provide specific strategies that individuals and organizations can use to protect confidentiality.
- Often, we find that we need to look outside of our discipline and profession for guidance on some ethics topics, as behavior analysis is still relatively new. Donner et al. (2008) have a four-article series in the "Focus on Ethics" section covering topics such as confidentiality as a primary focus for practitioners, enhancing our understanding of confidentiality in the contexts of positive ethics and principle-based ethics, and calling us to remember that ethical obligations must be applied flexibly in relation to the unique humans we serve.

Week 12

Check in—Whew! You are doing a great job working through some broad content that relates to many different areas of what we do. It can most definitely feel daunting and discouraging if we don't know where to start. Read through the scenario as you synthesize and pull together the content, and all the great reflections and contributions that you bring by engaging with this text. You are evolving and growing as you go.

Scenario—Read the scenario below, but don't look to label things as black and white in terms of right and wrong. That is the easy part. Use this scenario to practice all the nuanced considerations and skills you are learning as you develop your daily practice. Read with a different lens of risk, anticipation, caution, and action that could have—should have—been taken to minimize vulnerability and maintain confidentiality, dignity, and respect.

> *Sam has been a certified behavior analyst for 1 year and recently accepted her first supervisee accruing experience hours. Sam works for an organization with a great infrastructure for super-*

visory responsibility, data collection, and documentation, and she was very excited to support her new team member, Janice, with her coursework and experience hours. Sam and Janice followed the organization's guidelines, documented their activities, and made copies of the required forms for them to each keep.

One day, Janice found one of her supervision forms, which included feedback and things she needed to work on, in the copier, and she asked Sam about it. Sam apologized, saying, "Oops. Sorry! Must have left it there when I was making copies early this morning. My bad." Janice felt uneasy about how aloof Sam seemed, but figured it wasn't a big deal.

Throughout their supervisory time together, Janice observed a few other things that Sam did that didn't feel quite right. For example, on several occasions Janice overheard Sam discussing sensitive topics, including medication, financial responsibility, and co-pays, with families during pick up or drop off in the waiting room with other caregivers and clients present. A few times there were even job applicants and a delivery person in the waiting room. Sam also frequently took phone calls on speaker with supervisees and caregivers discussing specific details of their needs, and she left her office door wide open.

One afternoon, Janice observed Sam telling stories about a family's divorce-related argument during a session with another behavior analyst who has nothing to do with the case. Sam was clearly not seeking feedback or support from her coworker. It seemed, to Janice, that Sam was sharing the details of the incident just to share them. Janice consulted with Sam's supervisor, described the things she'd been observing, and shared her perspectives related to the importance of confidentiality and protecting patient dignity. The supervisor validated Janice's feedback and thanked her for sharing her observations.

- **VALUES CHECK**—As you were reading through, how did Sam's behavior and interactions make you feel? Did they seem respectful and aligned with our responsibility to maintain confidentiality and client dignity? Did Sam's priorities and focus seem aligned with her responsibility to protect her supervisees' confidentiality? Has this perspective informed your thoughts on values related to this topic?

- **IDENTIFY RISKS**—Circle the language that indicates a risk was present, and write down the resulting impact and harm related to all individuals involved.

  Can you think of other similar risks, given the observations made during this scenario, where additional training and feedback would be appropriate?

Be sure to consider specific aspects of the scope of risks or harms!

- **IMAGINE ANTECEDENT STRATEGIES**—Draw an arrow where antecedent strategies were missed or make a list, because there are many missing! How did these missing antecedent strategies impact the likelihood that the right behaviors would occur and increase the vulnerability of all involved?
- **IDENTIFY RED FLAGS**—Underline the red flags you found. What factors do you think could have contributed to Sam's ethical blindness and the occurrence of these red flags?
- **IMAGINE INSTANCES OF SOMEONE ADDRESSING THE ISSUE**—What do you think would be appropriate actions for the supervisor to take? Does it seem like Sam is ready to be a supervisor or has the training and resources to ensure that she is setting a good example for her supervisees in such a critical area? It would likely be good for Sam's supervisor to draw from tips on *Crucial Conversations* (Grenny et al., 2022) to outline her intention for the feedback. This should include the importance of Sam reflecting and refining her behavior to minimize future vulnerability to herself, and potential harm to others.
- **CONNECT**—When have you observed and experienced situations that are the same or similar to this before? What did you do, and what would you do differently now?

Taking Care of Business—This is a great topic for a team!! This is tough to tackle alone because it is so broad. Before closing this section and moving to the next, scan through your notes and your highlights, and take some time to draw on your mad self-reflection skills. Who will you connect with, and what changes will you evoke to expand your learning and impact moving forward? Start small, because starting is the only way to go somewhere, no matter how far the ultimate destination.

Take the next step to improve the care you take to manage information that is not yours and protect the dignity of others who need you. Remember, this is about daily practice and engaging in regular thought, observation, discussion, and action. You've got this, and we are right there with you!

Words of Inspiration—The essence of this section is the very old rule to treat others as you would expect to be treated—speak about them as you would want to be spoken about—and respect them as their loved ones do.

Weeks 13 & 14—Tending to Your Supervisory Practices

Week 13

Check in—We hope that the biweekly pair on confidentiality helped you identify things you are doing well and maybe a few areas for improvement. And hopefully you are making progress on the tasks you've been identifying each week. It's okay if you aren't getting to all of them; you'll have more time in a few weeks for catching up.

Focus for the Week—We focused on knowing about supervision requirements and strategies for building a strong foundation for the supervisory relationship in Weeks 7 & 8. Let's take a minute to revisit why supervision is so critical to the health and success of the profession of behavior analysis. It's no surprise to you that becoming an independently practicing behavior analyst requires the acquisition of knowledge and skills and then learning how to apply all of that to the service delivery setting. For those who are certified by a national or international body and/or practice in a licensure state, you likely had to complete a certain number of supervised hours.

Supervision is about consumer safety and about skill development. The behavior analysts who supervised you probably focused on providing direct training, observing you apply your skills and knowledge, and reviewing your permanent products, like reports and programs. They also should have actively worked with you to develop skills critical to successful and safe independent practice (e.g., assessment, program development and monitoring, case conceptualization, interpersonal communication, collaborative and therapeutic relationships, culturally responsive practices, training and performance management, time management and organization, personnel management, applied ethics).

They shaped your clinical and professional skills, and hopefully specific supervisory skills, that you now use in your clinical and supervisory practice. This shaping—this passing along of perspectives, skills, knowledge, and practices—happened regardless of your awareness. And you will similarly shape your trainees and supervisees. In this way, supervisors impact the profession of ABA-based service delivery far into the future because at least some of their supervisees and trainees will become supervisors passing along skills, perspectives, and practices, and those trainees will go on to become supervisors, and so on, and so forth, forever.

Now that we understand that supervisory practices are critical, let's briefly discuss what we mean when we say that as a supervisor, you need to design and evaluate your supervisory practices and systems. You already have all the skills needed to do this. You design effective service delivery practices and systems by doing the following: (1) practicing within the profession's scope of practice and your individual scope of competence; (2) developing trusting, collaborative relationships with clients, caregivers, and others; (3) implementing ethical, evidence-based assessment and intervention practices based on the principles of behavior analysis; (4) carefully and regularly evaluating the outcomes using accurate and valid data; and (5) adjusting when needed. See? So, all you need to do is implement steps 1 through 5 with your supervision practices!

Easy peasy! The full scope of high-quality supervisory skills is far too broad for us to cover in this book. And fortunately, we don't have to because there are many great books and articles that do a lovely job of outlining the needed skills and provide guidance—we've listed some in the "Resources to Check Out" section. So, we will focus our time on the ethics related to the overarching aspects of effective supervisory practices (e.g., taking a competence-based approach, using behavioral skills training (BST), implementing reinforcement-based practices, providing and documenting feedback) and evaluating the effects of your supervision.

1. **supervising within your individual scope of competence**
2. **developing trusting, collaborative relationships with supervisors and trainees**
3. **implementing ethical, evidence-based assessment, training, and performance management strategies based on the principles of behavior analysis**
4. **carefully and regularly evaluating the outcomes of your supervision using accurate and valid data**
5. **adjusting when needed**

Values Check—Alrighty, we suggest that you look over your values and give a big hug to the ones that most relate to this topic. Why? Because you might end up having some feels as you reflect on the supervision you received and the supervision you've been providing. And by feels, we mean the uncomfy ones that often produce avoidance responses. Use your values to keep you connected with forward-moving behavior, even when you feel frustrated or overwhelmed.

Risks & Benefits—Okay, so let's start by looking at close-in risks and then expanding our lens. Because supervision is critical to the delivery of ABA-based services, poor practice can translate to direct and real harm to technicians, clients, and caregivers (e.g., failing to train properly, not detecting when interventions are ineffective, missing the need to progress programming when a client is making gains). These risks can translate into lost time, money, and trust. Consumers and other onlookers may generalize the experience they have with one cruddy practitioner to be representative of the profession, and those perceptions can be barriers to accessing effective care for their loved ones.

As we already mentioned, poor supervision can impact our profession far into the future, as defective or underdeveloped repertoires are passed along. Failing to regularly evaluate the effects of your supervision means that any harms or risks will likely go undetected and continue to have negative impacts now and in the future, both because you may continue doing what you have always done, and because you will likely pass along that approach to others.

On the flip side, ensuring that you are implementing high-quality supervisory practices and engaging in continuous evaluation of those practices will benefit your supervisees, trainees, and clients and have positive, lasting impacts on the future of our profession. Regularly evaluating your supervision, like regularly evaluating your clinical programming for clients and your scope of competency, allows you to replicate the things that are effective and allows you to detect where you might need to improve your knowledge, skills, and overall approach to supervision.

Antecedent Strategies—The first, and a critically important, strategy to get ahead of issues related to ineffective supervision is to develop a trusting relationship with your trainees and supervisees and actively work to solicit feedback from them about your supervision. Ensure that you are using evidence-based strategies, like BST (Parsons et al., 2012), and identifying competencies that you can reference to systematically teach and measure skills based on pre-determined criteria.

You should also get in the habit of regularly reflecting on and evaluating your skills, including your scope of competence with clinical practices AND supervisor practices. It may be helpful to flip back to Weeks 3 & 4 and just reconnect with the content around scope of competence. Plan frequent, informal, close-in reflection and evaluation (e.g., weekly time to review how things are going), as well as regularly scheduled, structured evaluations (e.g., monthly/quarterly reviews of your supervisee's outcomes and feedback, as well as client and caregiver data and feedback). Gather data to evaluate the effects of your supervisory practices from multiple sources (e.g., feedback from team and caregivers; acquisition, maintenance, and generalization data from your clients; similar outcome data from your trainees and supervisees—including measures on the "soft skills"; and feedback on procedural integrity and interobserver agreement data).

Once you have data for several trainees or supervisees, reviewing and comparing that information can help you spot themes across individuals that might shed light on skills you need to develop or resources you need to refine or create. Having a mentor or trusted colleague to review your evaluations and talk though success and areas that may be susceptible to improvement can motivate you to get your regular evaluations completed and to plan any needed follow-up actions.

Red Flags—One of the biggest red flags that there could be some ethical missteps with your supervisory practices is the quality of the supervision you received. If the supervision you received was haphazard and poorly structured, or if your supervisor failed to regularly solicit your feedback, evaluate the outcomes of their supervision, and include express instruction on how to design and evaluate supervisory systems, that's a big ol' red flag. If you don't have a structured approach to delivering and evaluating your supervision, yeah, red flag. Other red flags are if you have had recurring or common performance issues with trainees and supervisees.

In the workplace, some telltale signs that issues could be brewing include trainees or supervisees frequently asking to switch supervisors, frequent deficiencies or deviances with supervision requirements, and a high rate of trainees or supervisees being terminated. Remember, identifying red flags is not about looking to place blame. Instead, it is about looking for environmental factors that may be impacting performance so that we can learn and improve.

Addressing—Addressing issues with your supervisory systems is quite simple really, but it requires self-observation and planning. We are lucky that there are many excellent articles, books, workbooks, podcasts, and continuing education opportunities to help us structure and evaluate our supervisory practices. But it can be a daunting task, so you'll need a plan, and you'll need to pace yourself. Don't worry, we put a bunch of great stuff for you to check out in the "Resources to Check Out" section.

If you find that your supervisory practices need some improvements, or that you haven't been consistently evaluating the outcomes of your supervision, don't forget to have an open and honest conversation with your current supervisees and trainees. Sure, it might be an uncomfortable conversation (Pssst: Red flag to check to see if you are likely to avoid this conversation—connect with your values, then lean in.), but it is important that you use this opportunity to demonstrate how to take accountability and learn from mistakes. You can clearly describe what has been lacking, what the risks are if you don't make some changes, and what you are going to do to improve things.

This is also a great opportunity to ask your trainees and supervisees for some feedback and ideas. If you have concerns with another behavior analyst's supervisory practices or the practices at your organization, get yourself comfy, think about how you can approach the conversation with compassion and common goals in mind, take a deep breath, and have that conversation.

Weekly Actions

MONDAY—THINK ABOUT IT

- When you think about your core values, which ones are most directly linked to ensuring that you have designed (or are using) high-quality supervisory practices and are regularly evaluating the outcomes of your supervision? Can you link one or more of your values to behaviors that can help you live those values in your supervisory practice?

- Get to your comfy place and spend a good 15–30 minutes thinking about your experiences as a supervisee and trainee. Can you remember your supervisors doing any of the following: using a structured approach to supervision, regularly soliciting your feedback, talking to you about their self-evaluation, teaching you how to design supervisory systems, or teaching you how to solicit feedback and evaluate the effects of your supervision?
- Spend a little time revising your approach to supervision. You may find that you are taking a structured approach to designing and evaluating your supervision, but maybe you only use feedback from your trainees or supervisees to evaluate, as opposed to gathering information from different sources (performance data, client data, caregiver satisfaction data). Maybe you find that you have a lot of areas for improvement. Wherever you are with your supervisory skills and practice, it's okay! Be where your feet are, and take a breath. Remember you'll be working on developing a plan over the rest of this week and next that you can roll out over the next several months.
- Jot down notes and thoughts:

TUESDAY—LOOK FOR IT

- Spend just a few minutes doing a temperature check on the general supervisory practices at your place of work. Do you have plenty of resources and support, or might there be some opportunities for improvement? Is there a commitment to high-quality supervision from your leaders, or does the company take more of a transactional approach to supervision (i.e., just another task to do, another box to check)?
- Is there a mechanism for you to collect and report data on the outcomes of your supervisory practices? Does your supervisor

or leader regularly check in on your supervision activities and how things are going?

- How successful are the trainees and supervisees in your workplace? Do they struggle to accurately follow processes and implement assessment or intervention procedures after being trained? Do you know what the pass rates are for the various certification exams individuals take?
- Are there indications that there might be issues with supervisors? For example, do trainees and supervisees transition from certain supervisors frequently? What is the frequency of reportable supervisory deviations? How frequently are there issues with trainees' applications for certification exams (i.e., how often are submitted hours rejected)?
- Jot down examples:

WEDNESDAY—TALK ABOUT IT

- Spend a few minutes chatting with your colleagues about some of your reflections and observations from Monday and Tuesday. If it is comfy for you all, invite your colleagues to share about their past experiences, and discuss similarities and differences. How do you all feel about providing supervision? Do you all have a shared understanding of how important high-quality supervision is? Together, can you identify the things that are going well and celebrate those? Are there a few mutually agreed-upon areas for improvement?
- Open a conversation with your supervisor or leaders in your organization. Share with them the things that are going well and things that you are grateful for. Maybe they provide excellent support for first-time supervisors or have a well-designed competency-based supervision curriculum.

  If you have detected areas for improvement for the organization or for your professional development, lean in and have that conversation in service of your values and commitment to the safety and health of the profession. You might start by saying something like, "I see how much this company values high-quality services and outcomes for clients. I wonder if we could chat a bit about how our supervisory practices might be tweaked to improve client services and outcomes even more?" Or, if you need some direct support to improve your supervisory practices, maybe you could lead with something like, "I have been spending a lot of time reflecting on the supervision I received and my supervisory skills, and I would love to have some support from you to help me improve (fill in the blank)."

- Use the information from Monday and Tuesday to open a conversation with your supervisees and trainees. Again, you can expressly highlight the bright spots and describe areas for improvement. If there are things that you are going to do differently, be honest with them.

  It's okay to be vulnerable and share that this is a learning opportunity for you and for them. Ask them for their feedback and ideas for improvement, and be specific. Avoid asking things like, "How is supervision going?" in favor of specific mands for information like, "What are three activities that have been really useful?" and "What is something we could add that would immediately make you more successful?"
- Jot down notes and thoughts from your conversations:

THURSDAY—ACT ON IT

- Make a list of the things you are crushing, and be specific! This list might come from your own reflection or praise from your supervisees or supervisor. Hang onto this list so that you can remind yourself of how great you are doing. As you make improvements, remember to come back and add to the list.
- Now, address red flags and risks you identified for yourself or your organization. Make a list of the specific things you think you can improve on related to your supervisory practices and the steps you take to evaluate the outcomes of your supervision. Maybe you've got a great structure for both, but you only evaluate the outcomes at the end of a formal supervision contract,

as opposed to at regular intervals (say monthly) throughout the supervisory relationship.

- Go back through your list and prioritize them based on things like ease of implementation, available resources to implement them, scope of the risk/issue they will address and improve, and how much they will improve your supervisees' and trainees' experiences and those of your clients.
- Jot down thoughts and tasks:

FRIDAY—REFLECT ON IT

- Honestly, we have asked a lot of you this week. Supervision activities likely take up a lot of your time each week, and you may be feeling overwhelmed. Spend some time just reflecting on the areas for celebration from the list you made yesterday.
- Take a few minutes to really reflect on the importance of well-designed supervision that is regularly evaluated and revised. What will our profession look like in 2 years, 5 years, or 10 years if we provide crappy supervision and teach others to provide crappy supervision? Really use your imagination and picture things in your mind. If using your mind's eye is challenging, maybe write it down or sketch it out.

  What will that look like for practitioners? For supervisees and trainees? For clients and caregivers? For us, it looks bleak. We imagine a zombie apocalypse with crappy supervisory practices infecting supervisees and trainees and harming clients and caregivers and the profession.
- Now imagine the future of your profession if we all commit to implementing well-designed supervision, evaluating our supervision regularly, and teaching others to do the same. How does that future look different? For practitioners? For supervisees and trainees? For clients and caregivers?
- Jot down something new you learned or something that surprised you:

Resources to Check Out

- Newer to your journey as a supervisor? Fraidlin and colleagues (2023) wrote an article focusing specifically on supporting newly certified individuals in their supervisory practices that is full of practical strategies to guide you on your way.
- The article by Turner and colleagues (2016) outlines steps to take that will make it easier to develop responsible supervisory practices.
- Kazemi and colleague's handbook, *Fieldwork and Supervision for Behavior Analysts: A Handbook* (2019), provides useful information and strategies starting with the trainee experience and moving through developing supervisory repertoires.
- The book on supervision and mentorship by LeBlanc, Sellers, and Ala'i (2020) will help you identify who you want to be as a supervisor and will support you as you design a competence-based and self-evaluative supervisory practice.
- The workbooks *The New Supervisor's Workbook: Success in the First Year of Supervision* (Sellers & LeBlanc, 2022) and *The Consulting Supervisor's Workbook: Supporting New Supervisors* (LeBlanc & Sellers, 2022) are all about how to take an active role in designing your supervisory practices such that they are individualized, effective, and continually evolving.

Week 14

Check in—Whew! Hopefully we did not give you nightmares with our dystopian supervision future. How are you doing? Are you feeling like you have an insurmountable mountain to climb? If so, take a deep breath, and remember that slow is smooth and smooth is fast. What the heck does that mean? Well, it means that no matter how many areas for improvement you identified, you can take a systematic approach to making adjustments and improvements over time. Don't rush things, just get started. Maybe you're feeling pretty good about things. If so, maybe this is a good time to check in with your colleagues at other organizations to see how they are doing and if you can lend a hand to facilitate improvements.

Scenario—You know the drill by now; spend some time reading the scenario and completing the prompts. Keep in mind that our ask of you is to focus on engaging in critical thinking about things like risks, benefits, values, and red flags to attend to, not just a surface evaluation of "ethical or not" and what standards might be compromised (although linking the scenarios to specific core principles and standards is a great start).

> *At Unicorn Behavior Services (UBS), Moshe has a caseload of 12 clients, and of the technicians on his team, three are accruing their supervised fieldwork experience hours. Moshe did not have the best supervision, which he did not realize until he began working at UBS. When he started, his clinical director, Amneh, asked him to engage in some structured self-reflection on his past experiences as a supervisee and trainee, and together, they were able to identify some skills that he was missing and map out a plan to build those skills.*
>
> *Moshe was appreciative for the opportunity to build his skills and make sure that he not only provided great supervision but also taught those skills to his supervisees and trainees. Amneh was impressed with how open Moshe was to engage in an evaluation of his past experiences and skills and was impressed with how quickly his skills developed.*
>
> *As she trained him on the supervision systems at UBS, which included a competency-based curriculum, supports for creating a collaborative supervisory relationship, and regular evaluations of supervisory outcomes using multiple data sources, she invited him to give her ideas for improvements. Moshe suggested that they track the frequency of supervisory issues (e.g., instances of a supervisor refusing to sign a monthly or final supervision form, reportable supervisory deviations, terminations of clinicians or team members for issues related to supervision).*
>
> *Amneh loved this idea, and she and Moshe worked to develop a data collection and review system. They were able to identify that there was one supervisor who had proportionally more trainees transfer to another supervisor as compared to other supervisors and a pattern of supervisors refusing to sign monthly forms due to trainees*

failing to implement feedback. With those data, Amneh was able to provide extra support to the supervisor in question and conduct an evaluation of the feedback skills of supervisors. This led to the creation of standardized training on feedback delivery, how to document feedback provided, and a feedback evaluation system for supervisors and their supervisees and trainees to systematically track performance with feedback-related skills over time.

- **VALUES CHECK**—Now that you have read the scenario, can you guess some of Amneh and Moshe's core values? How can you tell? Do they match your own values?
- **IMAGINE RISKS**—This is one of those scenarios where there might not be any obvious risks present. Can you think of things that could have happened differently and resulted in some risks of harm? If there are no risks/harms, based on the scenario as written, draw a circle over any areas where you could imagine the events playing out differently and inviting risks.

  Imagine the events or factors that would need to be present to invite the risks/harms. Jot down some indicators that you could look for in a similar situation.

  Be sure to consider specific aspects of the scope of risks or harms!
- **FIND ANTECEDENT STRATEGIES**—Go back through the scenario, and draw an arrow to indicate anything you think functioned as an antecedent strategy. What did Amneh, Moshe, and UBS do specifically to avoid risks of harm and to increase the likelihood of benefits occurring? Make a list of the antecedent strategies, and evaluate if they were effective at preventing things like risks or actual harms. How do you know? Could anything have been improved or added? If so, jot that down.
- **IMAGINE RED FLAGS**—Again, you might not be able to identify any red flags in this scenario. The idea is not to just hone your skills to identify red flags and risks that are occurring but to use your moral imagination to think about where they could pop up. If there were no red flags in the scenario, think about and write down a few that could show up in a similar scenario. Maybe you

can even think of some from your own experiences or from examples that your supervisors or colleagues have shared.

- **IMAGINE INSTANCES OF SOMEONE ADDRESSING THE ISSUE**—Since UBS, Amneh, and Moshe were pretty awesome in this scenario, there is probably nothing that needs to be addressed. However, based on your imaginings of risks and red flags, put an * next to a place where you think someone in the scenario could have taken action to address a potential issue. Write a brief description of the action that should be taken and how they would know if it was successful or not in the event of the imagined red flag or risk.
- **CONNECT**—Have you ever been in a situation with such well-developed supervisory practices and open feedback culture? If so, jot down the positive outcomes. If not, what were the issues (red flags and risks), and were they addressed? If not, what were the outcomes? If they were addressed, how (what strategies were implemented), and what were the outcomes?

Taking Care of Business—Use the rest of this week to really lean into taking a structured approach to outlining the tasks for improving your or your organization's supervisory practices, including how the outcomes of supervision are evaluated. Don't worry about getting started making improvements this week, unless there is an issue that is causing immediate harm. Instead, take your time. Effective supervision is a marathon, not a sprint. You'll also notice, if you look back, that we refer to supervisory "practice." Why? Because we will not always get it right, and we should always be working to improve our skills and processes related to supervision.

Go back to the list you started last week on Thursday. Schedule recurring events this week, if you can, to review and refine your list. Leverage your colleagues, supervisor, and mentor to give input into the tasks, implementation plan, and resources. And don't neglect that list of celebrations; those are just as important!!

Words of Inspiration—Listen, our work as supervisors is not easy. It requires that you attend to multiple intersecting contingencies, relationships, repertoires, and processes. You've got to think about the things, do the things, teach others to do the things, teach others to teach others to do the things, evaluate others' doing, and evaluate your doing. It is a lot. And you can do hard things. Connect back to your values, be kind to yourself, go slow and steady, and keep your eye on the horizon. And know that you have at least three colleagues cheering you on!

Weeks 15 & 16—Facililtating Your Awareness of Biases

Week 15

Check in—Be honest; how are you feeling? Is it getting easier to integrate looking for, considering, and talking about ethics throughout your days? At this point you might be feeling like you can't stop thinking and talking about ethics. Maybe you are finding yourself wishing for an ethics pause button. If so, that's totally normal. Feeling hyperfocused is part of the process of developing your daily practice. That noisy voice in your head will settle down as soon as your practice becomes more fluent and fully integrated into your routine and thought process. You are doing great!

Focus for the Week—It's no surprise to you that humans are complicated and that we all have many things going on at any given moment that influence our behavior. Factors like our mental and physical health, financial stability, relationship status, and caregiver roles all impact us and how we interact with others. Imagine you had an early morning argument with your significant other, spilled coffee on yourself as you got in the car, got stuck behind a slow driver, arrived to work 20 minutes late, stepped in a puddle walking into the building, and finally snapped at a technician who asked you to show them how to run a program. We've all been there. Our personal challenges can result in us behaving in ways toward our clients, supervisees, trainees, and others that are less than awesome.

As a behavior analyst, you understand we are, at any given moment, the sum total of all our experiences. These experiences can lead to us developing attitudes or associations about groups of things or people and characteristics. Makes sense from an evolutionary viewpoint, right? I see several people eat purple berries and they go about their daily tasks. I see others eat green berries, and they fall dead. Okay, purple berries—edible, safe, good; green berries—poisonous, unsafe, bad. That's a really helpful pattern to recognize that can inform my choice-making to help keep me alive.

But nowadays we are bombarded with information all the time. And sometimes, that information is presented in a particular way to send a specific message or have a specific impact on us, like with marketing or political campaigns. For example, if we are most often presented with depictions of purple people who are smart, capable, or kind and depictions of green people who are dumb, bumbling, or cruel, then we are likely to recognize a false pattern of purple people—good; green people—bad. The researchers at Project Implicit describe implicit bias as "an automatic reaction we have towards other people" (Project Implicit, n.d., para. 3). Our attitudes (or biases) and prejudices (opinions about someone that are not based on actual experience with that person) can negatively influence our behavior (covert and overt) toward others, even in ways that do not match our beliefs and values (Brownstein, 2019; FitzGerald & Hurst, 2017; Project Implicit, n.d.).

Alright, we didn't mean to be a downer. But the reality is that life is complicated, and we are complicated. So, if we are to behave ethically and in service of others, we've got to make space to acknowledge our challenges and biases and actively manage ourselves so that those challenges and biases don't negatively impact others. And since many of us are also in supervisory roles, we also need to support others to actively manage their biases and challenges.

These 2 weeks are best navigated with a heavy dose of self-compassion and require that you are willing to lean into some discomfort. Take a moment to self-reflect and reconnect with your values. Now do a scan of your bandwidth and stress level. If it's not a great time for this topic, no worries. We are proud of you for self-assessing what you need right now to be effective. The pages containing this ethics topic aren't going anywhere, but you can go forward and simply make a note to return to this content when you're ready. If you're ready to dive in, cool. Maybe take a deep breath or two as you get ready to tackle this topic over the next 2 weeks.

Values Check—It will probably be helpful for you to spend a good bit of time reviewing your values before you move though this biweekly pair. Some of the content up to this point might have made you feel a little uncomfortable. But with this topic, depending on your journey up to now, you may feel like your discomfort has leveled-up. Why? Well, the activities might help you uncover underlying assumptions that you have that stand in direct contrast to your values, which can be confusing and result in wanting to move away from this content.

So, sit with your values for a while and ask how they can serve you so that you keep pressing on with this topic. Maybe even go back to the values section in Chapter 3 just to reconnect with your values. You might even want to jot your values down here so they can serve as a prompt to keep leaning in.

Risks & Benefits—Okay, real talk. The idea is not that you are a bad person because you have implicit biases, prejudices, or life challenges. Quite the opposite. It simply means you are a human. The risks and benefits come from your willingness, or lack of it, to identify your implicit biases, prejudices, or life challenges and then choose to behave differently. Some of the risks related to failing to be aware of and attend to your personal factors, challenges, or biases include things like being disorganized, producing low-quality work, engaging in ineffective or harmful interpersonal communication, behaving unfairly or inequitably, and damaging relationships. Any of these possible undesirable risks can then lead to more severe risks of harm like ineffective programming, missed due dates that impact services, and emotional harm to others.

This example is offered without judgment—imagine that you just had a baby and are experiencing postpartum depression that is significantly impacting your sleep and memory. Imagine that you are responsible for administering medication, or entering clinical or billing data, and when you check the documentation, you see that you did not enter anything, and you cannot remember if you gave the meds or entered the data.

As with almost every ethics topic in this book, the primary benefit of actively and continually paying attention to these things is that you increase the chances of keeping your clients, caregivers, supervisees, trainees, and others safe. Actively managing personal factors that might negatively impact others allows you to take active steps to protect against those impacts. Yes, the process of noticing, acknowledging, and then taking action will likely bring up feelings of discomfort and even embarrassment. But another benefit of being proactive with these things is that you are modeling how to behave with integrity—how to do the difficult things. And, as you lean into this practice, as you make space for vulnerability in the service of others, you may find that your relationships strengthen and deepen, as this practice often increases our capacity for perspective-taking and compassion.

Antecedent Strategies—Similar to the strategies for scope of competence, discussed in Weeks 3 & 4, taking a proactive approach to this topic involves some self-reflection and evaluation. To begin to identify possible personal factors or challenges that could present a risk of harm to others, you might take time to make a list of the things going on in your life. For example, you might try writing down the following categories on pieces of paper: physical health, mental health, finances, relationships (familial, romantic, social), and work. Then, for each category, draw a line down the middle of the paper and list out facilitators (i.e., supports or things that are helping) on one side and barriers (i.e., things that are challenging or making things more difficult) on the other. Next, reflect on ways that the barriers might be showing up in your work and if there are any indicators that others are being impacted.

For this antecedent strategy to be impactful over time, you need to get in the habit of doing some version of this check-in regularly. It might be a good idea to put a recurring reminder on your calendar to do a structured check-in with yourself monthly. Then, if needed, you should develop a care plan to help minimize risks and address needs (e.g., accessing healthcare, managing your medication, getting adequate sleep, requesting temporarily modified work duties). Level up by adding a reminder to do a quick check-in daily. For example, put a sticky note on your work computer or as the first daily event in your calendar to just rate your current state and have some strategies to help mitigate any possible risks of your personal factors or challenges impacting others (e.g., deep breaths, doing some stretches, repeating a favorite affirmation, listening to music, taking a walk, contacting your therapist).

Go back to the self-care content in Chapter 3 and Week 9 (your first rest week) for more ideas! You'll also want to embed these types of check-ins with your clients, caregivers, and team members. You can develop a simple rating system using numbers, colors, or icons to get a quick temperature check on how people are doing, and then follow up with additional questions if needed and appropriate.

Proactive strategies to address implicit attitudes, biases, or prejudices toward others are a life-long journey. The first step in that journey requires that you be willing to admit that you have them. Honestly, saying it out loud to yourself or writing it down is a great first step. And as a reminder, *having* an implicit bias isn't necessarily a bad thing. We all have them; it just means you are human. What matters is what you do once you identify that you have them.

Connecting back to your values can be helpful to motivate you to continue working on this even when it's uncomfortable (and it will be uncomfortable if you're doing it right). You can access online tests that are designed to measure implicit attitudes toward and among different groups and characteristics. And here's the thing: You don't have to share what you're learning about yourself with anyone. But you're the only one with a front-row seat to your history and your private events. Only you can discover that you associate certain characteristics with certain attributes (e.g., skin color with intelligence or ability, body size with work ethic, race with trustworthiness, gender identity with ability to manage crisis behavior).

Only you can make space between the triggering of an implicit bias and how you then behave toward that person. This content needs to find its way into your supervisory practices, perhaps as a description of your journey, sharing some resources, and an invitation to your supervisees and trainees to engage in similar self-reflection.

Red Flags—The red flags related to this content are likely unique to each of us and can even change over time. However, one broad indicator of personal challenges posing risks in the professional setting is changes in our typical behavioral patterns, affect, and interactions with others. If we find that we're being short with people, more irritated by everyday things, more quickly infuriated, cutting corners with work that impact quality, or purposefully lacking transparency, these are all indicators that a personal or work stressor is impacting us in a way that could bleed over to those we are responsible for in a negative way.

A common red flag related to our biases creeping up in ways that pose risks to others is that when we make space to reflect on our interactions honestly and objectively with others, we find that we are often influenced by our thoughts and feelings, as opposed to facts, and that we may not be treating individuals equally. Other red flags can be evident in others' behavior—they avoid you, they're disengaged, there's an absence of any healthy discourse of feedback, and there's a presence of interpersonal conflict. You can also watch for these red flags to help you detect that other people's personal factors or biases are impacting their behavior.

Addressing—If you find that your personal factors, challenges, attitudes, or biases are negatively impacting others, the good news is that you can immediately do something about it. You can engage in honest self-reflection, evaluation, and care-plan development. Remember to have grace and be kind to yourself as you work to address this ethics content. You might need to recruit the support of others to assist you as you address things. For example, you might need to talk to your supervisor or employer to develop, implement, and evaluate a plan at work that might include things like proactively planning out tasks and due dates to provide extra time; modifying your work activities; accessing related health benefits; engaging in regular check-ins; or having an emergency plan to minimize disrupting services to clients, supervisees, or trainees.

If these things have negatively impacted those you work with, you might need to deliver an honest apology that describes the issue/harm, how you perceive that it impacted the other person, and what you're going to do to ensure it doesn't happen again. For example, if you discover that an implicit bias has resulted in inequity in how you give access to opportunities at work, you might say something like: "Annalise, I want to acknowledge that I have not been fair and equitable in how I have given opportunities to work on certain projects and to attend conferences. That likely made you feel confused and undervalued, as your work is excellent. I have developed an objective rubric and tracking system that I will use from now on to make sure that opportunities are matched to performance and that I am rotating across all team members equally. I'd love to hear your thoughts and get feedback from you as I implement this new system so that I can make improvements."

Similarly, if you notice that someone else's biases or prejudices might be impacting their behavior toward others in the workplace, you might need to have an honest and compassionate conversation about this topic.

Weekly Actions

MONDAY—THINK ABOUT IT

- Just take a moment to check in with how you feel after reading about this ethics topic. Are you feeling uncomfortable? Defensive? Anxious? All those feelings are totally valid. ☕ Get yourself in your comfy space, and try to just observe and describe your covert behavior and feelings. You might try using this prompt to say or write down your thoughts or feelings: "As I observe my thoughts/feelings, I notice that I am thinking/feeling (fill in the blank)."
- Ask yourself how often you've engaged with these topics and in what contexts. Maybe you have done a lot of work related to this topic. If so, that's wonderful—keep going, and remember to be kind to yourself. This could be quite new to you, and if it is, take it slow. You might want to pause here and come back to this content in a few weeks. Remember that this ethics topic requires you to commit to active, ongoing practice.
- Think about your personal challenges and biases. Don't judge them, and don't label them as bad or problematic. Don't say *It's bad that I (fill in the blank)* or *My (fill in the blank) makes me weak/a bad person/a bad behavior analyst/a lousy supervisor*. Just spend some time making space to list out the things in life that have been difficult and biases that you have noticed coming up for you.
- How do your values align or show up related to taking an active role in identifying your stressors, challenges, and biases? How can you use your values to help you lean into a commitment to continually engage with this topic and talk about this topic with others?
- What is your learning history related to personal factors or challenges? For example, did you (like at least one of the authors) grow up in a family that did not talk openly about personal struggles and actively hid them? Do you suspect or know that your

past supervisors or employers behaved in ways that were influenced by personal challenges or biases?

- Can you identify risks or harms that you experienced because others did not manage their personal factors, challenges, or biases? How did that impact you?
- Can you identify risks or harms that occurred because your personal factors, challenges, or biases influenced your behavior?
- Think about how the risks and harms might be similar or different across clients, caregivers, supervisees, and trainees.
- If someone asked you right now, "What are your personal challenges (now or in the recent past) or biases?" what would you say?
- Jot down notes and thoughts:

TUESDAY—LOOK FOR IT

- Can you identify instances where others' personal factors, challenges, or biases have or are negatively impacting people in your work setting? Were/are there red flags, and if so, can you describe them? What were/are the risks (i.e., how did the person's behavior negatively impact others)?
- What is the culture in your workplace related to talking about personal factors, challenges, and biases? Do team members generally feel safe enough to talk about these things?
- What sort of focus does your workplace have related to areas like diversity, equity, inclusion, and accessibility?
- What is the culture in your workplace related to self-care? For example, are there regular check-ins with team members, are they encouraged to take breaks and scheduled time off, are self-care and stress-mitigating strategies (e.g., job crafting, regular high-quality supervision, the opportunity to give feedback and nominate ideas to improve the workplace) encouraged and implemented?
- Jot down examples:

WEDNESDAY—TALK ABOUT IT

- Start a conversation with a colleague or supervisor about this topic. How you start the conversation will depend on the culture of your workplace. You could focus on talking about how this topic is important and how it's linked to ethics. You could chat about strategies you each use or would like to try related to this topic. You could review any antecedent strategies in place, discuss how they are working, and brainstorm a few new ideas. Maybe talk about how the risks and harms related to this topic might show up differently for clients versus supervisees and trainees.
- Reach out to a trusted mentor or colleague/friend, especially if you know someone familiar with this topic. Consider asking what they do to attend to their personal factors, challenges, and biases. Maybe ask them if this topic ever makes them uncomfortable, and if so, how they address those feelings. If you are comfortable, share any concerns you have with this topic, including those related to how to talk to others about it.
- Bring up this topic with your supervisees or trainees. Start by focusing on how this is related to ethics, how it's likely to come up for all of us at some point in our careers, and what the risks are if we don't actively attend to this topic. Depending on your relationship with your supervisees and trainees, you might share some of the activities described in the antecedent section or those in the "Resources to Check Out" section. It's important that you acknowledge the inherent power imbalance between you and your supervisees and trainees, and that you do not require them to share the results of any of the activities. You need to meet your supervisees and trainees wherever they are on their journey.
- Jot down notes and thoughts from your conversations:

THURSDAY—ACT ON IT

- Assess your past life factors and stressors following the steps listed in the "Antecedent Strategies" section. Why focus on past events? Looking back honestly and objectively can allow you to identify how you responded to those events and to look for risks and red flags a little more easily than evaluating your current situation.

- Assess your current life factors and challenges, focusing on what introduces additional challenges to your daily activities. It may be helpful to think about factors and stressors in three different buckets: (1) long-standing (e.g., chronic health needs, mental health needs), (2) recurring and predictable (e.g., holidays, performance evaluations, allergy season), and (3) recurring and unpredictable (relationship difficulties, illness, unexpected financial costs).
- Make a care plan that focuses on managing your wellness and self-care across those three buckets. In addition to mapping out proactive strategies and resources, spend some time identifying your red flags (e.g., changes in your behavior or the environment) that are reliable precursors to increased stressors or challenges.
- Specific to implicit biases and prejudices, consider taking some of the freely available online assessments (see the "Resources to Check Out" section). In completing these assessments, you may have uncomfortable thoughts and feelings. It can be jarring to discover that we have biases or prejudices that don't match our beliefs or values. It's important to make space for them, reflect on them, and look for examples in your behavior where your biases have shown up.

  The next step is to commit to the practice of identifying, in the moment, where biases show up for you moving forward and to make an active choice to behave contrary to those biases. We recommend that you think about accessing them in small doses, rather than trying to take on several all at once. For example, you may complete one every other week, scheduling in time to reflect on the results every few days.
- Prioritize accessing some professional development activities on these topics. You might identify a few articles or on-demand webinars that you can tackle on your own, with a friend, or in a group. At the next conference you attend, set a target goal for attending a certain number of talks focusing on these topics. Identify some podcasts or books on these topics. Identify a few trusted colleagues and schedule a recurring time to discuss

these topics—and make it fun by chatting while you all engage in an activity (e.g., hiking, cooking, or sharing a meal).

- Add reminders in your environment. Add 15 minutes weekly to simply sit in quiet reflection. Add recurring monthly or quarterly reminders to do a structured review of your life factors, challenges, and stressors. Add "check-in" as a standing item on your agenda.
- Revisit your discussions with those in your workplace, and make an action plan to implement any of the ideas you brainstormed (e.g., develop a quick visual check-in scale or system for team members).
- Jot down thoughts and tasks:

FRIDAY—REFLECT ON IT

- Listen, these topics are heavy. So, please spend some time taking care of yourself. Reflect on how interacting with these ethics topics is really about placing clients, supervisees, and trainees at the forefront of your work efforts to keep them safe. And that is commendable.
- Reflect on the thoughts and feelings that came up for you this week. Did you notice any instances where you felt defensive or wanted to move away from this content? If so, that is completely understandable. Maybe reflect on what you can do to stay engaged with these topics.
- Reflect on what an incredible person you are. Yeah, that's right! It's helpful to add in some time to reflect on your accomplishments, things you have overcome, and the good that you do in the world. So, take a deep breath and just sit quietly with yourself, content in the knowledge that you are doing amazing things, hard things, worthwhile things.
- Jot down something new you learned or something that surprised you:

Resources to Check Out

- The Behavioral Healthcare Providers' *Recent Life Changes Questionnaire* (RLCQ; 2016) is an online tool that you can use to evaluate stress related to your "life change events" and reflect on the effect they may be having on you.
- Homewood Health's *Building a Self-Care Plan* (n.d.) is a lovely document that walks you through building a self-care plan. It takes you through a step-by-step process that includes evaluating your coping skills, pinpointing self-care needs, reflecting on roadblocks and targets for improvement, and finally creating your self-care plan. It also has a wealth of ideas for self-care activities you might try.
- Another resource for developing a self-care plan to check out is available through the University of Buffalo School of Social Work; it walks you through seven simple steps to reflect and plan (Butler, n.d.).
- Stephen Warley provides some thought-provoking questions aimed at helping you get to know yourself better and then gives you a model by providing his own personal self-assessment. Warley (n.d.): *My Personal Self-Assessment.*
- Taking the Project Implicit (2011) IATs is a good place to start uncovering your implicit biases related to race, gender, sexual orientation, and a variety of other culturally relevant topics.
- Dr. Kristin Neff is one of the founding scholars and researchers on self-compassion. Check out the online self-compassion assessment to get started. When you are ready to take action, you can also find a variety of helpful activities and exercises to support your practice (Neff, n.d.).

Week 16

Check in—Hey there. How are you doing? Maybe you're reading this the week after completing the first week of this topic. Maybe you needed a break between the first and second week. Either way, we're glad you're here to continue your practice of infusing every day with a focus on ethics, and specifically on attending to how our personal factors, challenges, and biases show up and can influence our behavior in the workplace. We hope that the reading and activities from the first week were helpful, wherever you are on your journey with this complicated ethics topic.

This week will hopefully give you some breathing space to apply the things we covered last week and get started mapping out and completing related tasks to remain connected with the practice of leaning into these topics. But remember, this is about the progress made through continued practice, not about achieving a final goal. Slow and steady is the best approach here. The words on these pages will always be here for you to return to, and to support you in your practice.

Scenario—Read the scenario and complete the prompts. Remember, the goal is not to identify the specific standard from an ethics code that was violated, although that is a great activity! The purpose is to think critically about *any* situation and see the ethics-related content and actions.

> *Brian is a recently hired clinical supervisor in a multidiscipline behavioral care company in a large city serving a diverse population with an equally diverse team of professionals. He oversees other behavior analysts, speech-language pathologists, psychologists, and occupational therapists to support them in managing their cases and completing operational and administrative tasks. He is responsible for their performance evaluations and for the subcommittees in the clinical department, including assigning tasks like presenting at conferences and participating in research projects.*
>
> *Brian is White, cisgender, straight, and male. Before this job, Brian spent his entire life and career in a small town in the Midwest of the United States with a very homogeneous population. In his first few months, he has enjoyed supporting some of his team members and is developing good relationships with some of them. For others on his team, he finds that he worries about their skill sets, especially related to professionalism and interpersonal communication skills.*
>
> *For example, several of the team members have hairstyles and some wear head garments that he personally feels are unprofessional and could inhibit developing therapeutic relationships with clients. Some of these team members often talk in loud voices and use a lot*

of animated hand gestures that could be seen as argumentative or off-putting to clients. He worries about assigning these team members to high-profile clients or giving them assignments that include public speaking or representing the company.

He has recently noticed that some of these team members have been less engaged in group meetings. Performance evaluations are coming up soon, and he is sure that he will score several of them low, not because of their work performance, but because of these other unprofessional behaviors. When he compares their work performance to others on the team, they are equal or, in some instances, superior to the other team members.

He feels badly that they will receive low scores, as he knows that they will not be eligible for raises, bonuses, or promotion opportunities. He also feels sad that their previous supervisor did not address these issues. However, as their boss, he feels it is his responsibility to address their unprofessionalism and help them build their related skills. His supervisor offered to collaborate on the upcoming performance reviews, but Brian feels confident that he can handle the reviews.

- **VALUES CHECK**—After reading the scenario, how does Brian's behavior and approach to his team sit with you? Which of your values seem most related to this scenario?
- **IDENTIFY RISKS**—Circle language indicating a risk that was present, and then write down the risks/resulting harms.

  Did Brian identify the same risks, or any risks? Did Brian take action to minimize or avoid the risk? Are there other risks or actions you can think of that could be related to this situation?

  Be sure to consider specific aspects of the scope of risks or harms!
- **FIND ANTECEDENT STRATEGIES**—Go back through the scenario and draw an arrow to indicate anything you think functioned as an antecedent strategy, or places where an antecedent strategy could have helped. Make a list of any antecedent strategies from the scenario and evaluate if they were effective at

preventing things like risks or actual harms. How do you know? Could anything have been improved or added? Now make a list of the antecedent strategies that you think could/should have been implemented, and briefly describe what it would look like to implement them and how you would know if they were effective.

- **IDENTIFY RED FLAGS**—Underline any red flags that you find, even if Brian didn't notice them. If he didn't notice them, write down what factors may have contributed to his ethical blindness (e.g., time constraints, lack of knowledge, authority, common practice, perceiving that there was no risk, lack of oversight). Can you think of any others?
- **IMAGINE INSTANCES OF SOMEONE ADDRESSING THE ISSUE**—Not all scenarios will include a description of someone addressing the concern or issue. Put an * next to a place where you think someone (e.g., Brian, a team member, or his supervisor) in the scenario could have taken action to address the issue. Describe the action that should be taken and how they would know if it was successful or not.
- **CONNECT**—Reread this scenario and ask yourself, *What would I need to change to make this a clear ethics violation?* and *What would I need to change to make this example ethical?* Then ask yourself *Is there anything that Brian should do differently next time to improve things?* In other words, what would this scenario have looked like if Brian actively attended to opportunities to his own personal biases? Have you ever been in a similar situation or known someone who was? What was the outcome? What would you do differently, or recommend the other person do differently, now?

Taking Care of Business—Alright, the first thing we want you to do is take a deep breath. Maybe you plowed through the first week of this content and you are ready to use this second week to review, revise, and get an action plan going. If so, sweet! Get after it! Maybe you needed to take a break during or after the first week of content. If so, that is perfectly fine. We are proud of your work, and we invite you to take this week to just set your intentions to come back to this content and any tasks you identified when you are ready.

Words of Inspiration—Remember, having biases doesn't make you a bad person; it makes you human. We all have our own stories, and we can choose to notice and change our behavior when they are impacting our work. Be kind to yourself. Practice self-compassion. This journey requires stamina, so proceed along your path gently. We need you for the long haul.

Weeks 17 & 18—Resting & Reconnecting

Week 17—Time to Rest & Take Care of Yourself

You've made it through three more biweekly pairs and likely some challenging content. Yay, you! That means that it's time to take a well-earned break. Yep, that's right! Do your best to push pause on that big ol' brain of yours and just rest and recharge. Take a beat to check in with yourself. How are you doing? Is anything weighing heavily on your mind? If so, try writing your thoughts down and giving yourself permission to do nothing more than gently observe related thoughts for the remainder of the week. Maybe you are feeling energized? That's super cool too! However you come into this rest week should inform your daily intentions and practices.

Monday–Sunday Self-Care & Wellness Intentions

Take some time on Monday to set your intentions to engage in self-care and attend to your wellness. Flip back to Week 9, when you took your first rest week, and review the prompts, resources, and any notes you jotted down. Which things worked, and which ones not so much? Maybe there are some activities that you've integrated into your daily or weekly practice, and if so, keep those up! Maybe there are some new things you'd like to try this week. Either way, this week is for you!

Monday Self-Care & Wellness Intentions

Tuesday Self-Care & Wellness Intentions

Wednesday Self-Care & Wellness Intentions

Thursday Self-Care & Wellness Intentions

Friday Self-Care & Wellness Intentions

Saturday Self-Care & Wellness Intentions

Sunday Self-Care & Wellness Intentions

Ideas for Regularly Scheduled, Recurring Self-Care & Wellness Practices

Just like you did in Week 9, take some time midweek and on Sunday to jot down notes about anything you tried that you really enjoyed or things that were meh. Then, as you move into next week, see if you can add the beneficial activities or practices into your daily, weekly, or monthly plan of self-care.

Week 18—Time to Reconnect With Your Daily Practice

You are making such great progress! Hopefully your rest week included lots of recharge time and no ethics time. Just like in Week 10, we want you to take this week to ease back into your daily practice, but mostly we want you to take time to get caught up. We know that we ask a lot of you, and you are probably identifying even more tasks. For this reconnect week, and those that follow, you'll just take this time to remind yourself how awesome you are, get back to thinking about ethics, complete tasks, and plan out future tasks. We won't schedule it day by day, but we have provided some prompts for you to consider throughout the week.

Review & Reflect— Check in with yourself. How are you feeling about the ethics topics we've covered so far? Are you feeling any shifts in how you think and talk about ethics? Are you finding that keeping ethics in the forefront of your practice is more fluent?

Celebrate— Look back at the tasks that you have accomplished, and give yourself some acknowledgement. Maybe you've improved your practice or positively impacted resources or activities in your workplace. Maybe you leaned into a difficult conversation. Heck, maybe your success is that you made time to care for yourself last week and are making slow progress toward embedding recurring self-care and wellness activities into your routine. We are confident that you are making progress, and since it's so easy to focus on all the things we haven't yet done, we want you to just spend a few minutes celebrating the things you have accomplished.

Refine, Plan, & Act— Look back at your list of tasks and ideas that you haven't been able to get to, and identify those that you are sure you can accomplish this week. And listen up, if this week is tough and you know you won't be able to get to tasks, that's okay. Maybe you need to push this out 1 or 2 weeks, and that is perfectly fine. For the tasks you know you can't get to, get them on your calendar for the future. It's okay if you've got to bump them around, but putting them on your calendar increases the likelihood that you'll work on them in the future.

Tasks for This Week:

______

______

______

______

Tasks to Find a Cozy Spot on Your Calendar:

______

______

______

______

______

Have you found yourself regularly pushing out tasks or telling yourself that you're going to skip a section and come back later? If so, that is totally okay—we've been prompting it! But make sure to include some self-checks along the way. Are there patterns in what you are pushing out (e.g., tasks involving difficult conversations) and similarities in the topics you are skipping?

If so, why do you think that is? Is it maybe a series of topics that you're not in the right place to address? Are you avoiding the tasks, conversations, or topics because they make you uncomfortable? Are you avoiding the topic because you're worried you might have done something wrong in the past? Reflecting on these patterns and being aware of them are great ways to help you on your path to engaging in daily ethics practice. Make sure you're holding yourself accountable (you might even want to find a colleague or mentor to help you out with this) and creating a plan to come back to it later.

Look Ahead— Don't forget to scope out the coming biweekly pairs so you know what to expect over the next several weeks. Doing so can help you plan to make the most of your time with the different topics. For example, you might want to find related training topics or policies to have handy, or get some additional time with a mentor or your supervisor in anticipation of some needed discussions or planning activities. Maybe one of the topics is going to be trigger for you, or maybe now is not a good time for a topic. Looking ahead allows you to be proactive, maximize your time, and protect yourself. Sometimes self-care looks like saying no, or not right now.

Inspiration to Reconnect— As you lean back into your daily ethics practice, know that your efforts are making a difference. There may be times when you feel overwhelmed, and in those moments, remember to breathe and connect back to your values.

Weeks 19 & 20—Addressing Your Own Possible Misconduct & Possible Misconduct of Others

Week 19

Check in—Did you take time to rest and reconnect the last 2 weeks? We hope you're refreshed and ready to continue! You have worked through a lot of ethics content to get to this point, and you've engaged in a lot of self-reflection (some of which may have been a little uncomfortable). Amazing! You've probably also had a few difficult conversations as you've been addressing ethics topics and leaning into your daily practice. Let's keep this momentum going!

Focus for the Week— We know that the title of this section might have resulted in some apprehension about starting this biweekly pair. But the point of this section is not to make you feel like you're unethical or a bad person. Or to equip you to be the ethics police. Instead, the purpose of this section is to prepare you for times when you might identify that you've made a misstep in your practice, or you've identified a misstep made by someone else, and to help you evaluate what to do next. Let's start there.

If you are reading this and thinking *Oh, I've never really dealt with an ethical dilemma* or *I haven't had any missteps that I'm aware of*, let's pause. One thing we (the authors) have learned in our practice is that if you find yourself responding in this way, you likely need to take a closer look. We face ethical dilemmas all the time in our practice, and many times we can't even identify them until after they have happened. Being able to detect when we are at a critical decision-making point or supporting someone with making a critical decision is so important. And guess what?!? You are preparing for that now by reading this book.

What you've probably found in the last 18 weeks is that you've encountered a lot of ethical dilemmas or the potential for them in your practice and may not have realized it at the time. Believe us, all three of us have been there. The focus of this biweekly pair is what to do when you identify that you or someone else has engaged in potential misconduct, including steps to take to analyze and address the situation.

Values Check—What values came to mind as you were reading the focus for this week? Did anything jump out? What values have you identified that have driven you to do this work? Ethics isn't any easy topic, and taking on the daunting task and putting in the hard work to work this into your daily practice means that you have likely identified some values that motivate you to do this. We have a hunch that those values align with the values that you might identify for this topic. But take a moment, and look back at your values list and identify what values are leading you to lean into this topic and address ethical dilemmas.

Risks & Benefits—Let's split the risks and benefits into two categories: (1) addressing your own behavior and (2) addressing the behavior of someone else. We will use these categories for the rest of this topic.

Your Behavior & Risks—There are quite a few risks of not addressing your own possible misconduct. First, serious violations of the law or rights of your clients, trainees, supervisees, or colleagues, or anything that results in significant harm to others (e.g., abuse or neglect, physical harm, fraud), is a serious matter and needs to be promptly addressed. Not addressing it may result in further harm to those involved and more significant impacts to your licensure or certification status. We aren't trying to scare you; we just want you to be aware of the seriousness of failing to address your own missteps right away. Second, even if there was no significant harm, not addressing potential misconduct can put you at risk for continuing to engage in the concerning or problematic behavior.

For example, think about transitioning clients. What if you do not complete all the necessary steps for a smooth transition (e.g., you didn't create a transition plan to share with the next behavior analyst) for a client? You might not even be aware that you missed critical steps, so you'll continue to miss those steps for future clients. You are probably also modeling this behavior for your supervisees and trainees. And you are continuing to impact the progress made by the clients you are transitioning because they likely are stalling in progress or even potentially regressing while the new behavior analyst is taking over the case.

Also keep in mind that you may have a requirement to report your own potential misconduct to your organization, licensure board, or certification body, and not reporting may result in impacts to your status with these relevant bodies. Remember that those organizations are here to protect consumers, practitioners, and the profession, and we have a duty to keep them informed of things that might or will impact our ability to practice. Again, not trying to be dramatic or scary; we just want to remind you that as a professional, you have these responsibilities.

Your Behavior & Benefits—What are the benefits for addressing potential misconduct in your own practice? You might be able to fix the situation yourself and avoid future issues. If not, at least you'll be able to immediately reach out for help to address the situation before it gets worse. Acknowledging and addressing your own behavior will probably help you identify similar situations in the future, and you'll be better prepared to address them and prevent any potential misconduct.

And along those lines, what are the benefits of notifying your organization, licensure board, or certification body of the potential misconduct? When you submit a self-report, you can share the steps you have taken to resolve the situation and steps you have taken to prevent or address similar situations in the future. You can demonstrate that you have taken action after becoming aware of the issue (rather than not responding or not being aware of it).

And we hear you, it probably doesn't sound comfortable to say, "Hey, I think I made a mistake and I want to let you know about it," especially to a licensure board or certification body. But think about it this way—YOU are taking the action, sharing the information that led to the decisions you made, and sharing the steps you've taken to move forward and do better in the future. Yes, there may be action steps that you are required to take, or in more severe situations, there might be action taken against your licensure or certification, but again, you have taken the steps to demonstrate that you are motivated to do better and taking action to do so.

Others' Behavior & Risks—Okay, now let's talk about addressing the potential misconduct of others. The risks are the same in a lot of ways—if you elect to not address it, they are likely to continue engaging in the same behavior, putting their clients, supervisees, trainees, and colleagues at risk. And, unless the issue is very grave (e.g., violation of the law) or addressing it directly with the individual places you or others at increased risk of harm, failing to talk to the individual first (i.e., moving straight to submitting a formal complaint) will likely result in the risky or potentially problematic behavior continuing for quite some time, as the complaint processes are often lengthy.

Others' Behavior & Benefits—The benefits are also very similar to addressing your own potential misconduct. You might be identifying something that the person was not aware of or that they were struggling with addressing, allowing them the opportunity to change their behavior and work to resolve the current situation (if possible). By discussing the situation directly with the individual, you might also find that it does not actually rise to the level of needing to be reported to any of the relevant bodies (e.g., organization, licensure board, certification body), or you might even prompt them to report the incident themselves.

Antecedent Strategies—The first recommendation we'll offer is to familiarize yourself with the relevant ethics requirements for all the relevant bodies (e.g., organization, licensure board, certification body). This includes being aware of requirements related to reporting potential misconduct of yourself and others. You can incorporate discussing and thinking about ethics into your everyday practice (as you are as you work through these biweekly pairs!). You can also prepare for addressing ethical dilemmas by accessing resources that walk you through the steps to take when you identify a potential ethical dilemma or instance of misconduct.

The *Ethics Code for Behavior Analysts* (BACB, 2020a) provides one model for addressing ethical dilemmas comprised of 11 steps. Rosenberg and Schwartz (2019) provide a six-step decision-making model in their article. Check out the resources at the end of this week for more books and articles. You might also consider reviewing some of the resources we mentioned earlier for having difficult conversations (and listed below). You might even set up a time to write some scripts to start the conversation (with yourself and with others) and regularly practice these skills so that when the situation arises, you are less likely to run for the hills. Instead, practicing with scripts will help you lean in, and you'll be able to think on your feet and successfully respond in the moment.

Red Flags—An indicator that you might be at risk is if you identified that you have a history of avoiding acknowledging these situations when they occur. Maybe you identified that the amount of "uncomfortable" that you feel when identifying potential misconduct in your own behavior or someone else's overpowers your dedication to the obligation to make others aware of the situation and address it appropriately. Which, by the way, is a pretty normal reaction. Or maybe you find yourself saying "Well, no one is going to report me" or "No one is going to find out," or "There was no harm, so this does not need to be addressed." Or maybe you find yourself avoiding the conversation to address potential misconduct with someone else because you feel like it won't be successful, might hurt their feelings, or might make them upset. These are all red flags that are good to note and reflect on as you move through the rest of this content. These red flags are calling you to lean into this topic.

Addressing—So what should you do when you find that you need to address your own ethical misconduct? First, you might find it helpful to review some resources on addressing ethical dilemmas. The "Resources to Check Out" section at the end of this week has some helpful resources to support you. Then, you must consider the risk of harm or potential risk of harm to those involved and what steps you should take. There are some questions to consider.

- *Can the situation be rectified?*
- *Have you notified your supervisor and discussed the situation?*
- *What are the requirements related to your certification or licensure when you engage in behavior that may have potentially violated the ethics requirements specific to that organization?*
- *Do you need to self-report the event to the licensure or certification body?*
- *Were any laws potentially broken?*
- *Do you need to report as a mandated reporter?*

Ok. Let's pause here again. Those were a lot of questions, and you're likely feeling a little stressed thinking about going through this list of questions. But here's the thing. We must make sure we are protecting our clients, supervisees, trainees, and colleagues. And we must strive to engage in ethical behavior and decision-making. And when we don't, we must take the next steps to rectify the situation and put things (e.g., policies and procedures, resources, checklists) in place so we can avoid making the same decision in the future.

You didn't choose this career because you don't care about people. You did it because you likely really care about the people you serve, want to have a positive impact on those around you, and want to be the best behavior analyst you can be. And be kind to yourself. These situations are tough. And if you are so hard on yourself each time you make a decision that might not have been the best decision in the moment, this is going to be a tough ride. So, show yourself some grace and compassion, and commit to continuing to learn and grow and do better.

Now let's take this to our second category—someone else's behavior. What if you become aware of a colleague, peer, supervisee, trainee, or supervisor engaging in potential misconduct? Start by engaging in some smart questioning.

- *How will you approach this situation?*
- *Is your initial reaction that you want to tell them what they did wrong?*
- *Or report them to the relevant licensure or certification body?*
- *Or is your initial reaction that you want to avoid the situation and pretend like you are not aware of it?*
- *Or maybe it was somewhere in between?*
- *Maybe your reaction was something like "I know I need to address this, but I'm not sure what to do next."*

Take a moment and reflect on (or write down) what your initial reaction was. Why do you think that was how you reacted? Does who you imagine as the person engaging in the potential misconduct change how you initially react? Make note of these thoughts, and this might be something to take some time to reflect on again.

You might even take some time to reflect on any situations in the past where you've become aware of potential misconduct of someone else. How did you respond? Did it match your initial reaction when presented with the questions above? So, what might you do next? The questions from addressing your own potential misconduct are likely the same. But we suggest you pause and add one step. Were you present for the situation that resulted in the potential misconduct? Have you had a chance to chat with the person involved and get their perspective?

The first step we suggest you take is a conversation to get a little more information. Why did they make the choice or choices they made? Was there information or context that you were not aware of? Were they aware that their behavior was potentially (or actually) problematic? You might consider prepping some questions before you chat with the person, or even practicing the conversation with a supervisor, mentor, or colleague. And remember, approach the conversation with kindness and an openness to discussion. It's likely to be unsuccessful if you start it off accusing someone of misconduct. We've included some resources at the end of this week to support having these difficult conversations.

Weekly Actions

MONDAY—THINK ABOUT IT

- How do your values support you in addressing your own potential misconduct and the potential misconduct of others?
- What is your learning history related to addressing these situations? Were you taught how to address your own potential misconduct and the misconduct of others?
- Has anyone addressed your potential misconduct with you? If so, how did they do it? How did it make you feel? Are there things you think they could have done differently to make it a less scary/stressful experience?
- Can you identify any instances when you were aware of your potential misconduct or the potential misconduct of others and did not address it? Take some time to reflect on (or write down) reasons why you did not address it.
- Can you identify any instance when you addressed possible misconduct (either your own or someone else's)? How did it go?

- Jot down notes and thoughts:

TUESDAY—LOOK FOR IT

- Can you identify some examples of how addressing potential misconduct of yourself or others comes up in the workplace? Are there antecedent strategies in place to prepare you and others for addressing these situations?
- Can you identify some examples of where it could be beneficial to add resources and support to ensure that you and your colleagues are addressing potential misconduct?
- Does your organization have an ethics compliance officer, ethics committee, or other similar avenue for seeking support for addressing potential misconduct?
- Does your organization collect data on ethical dilemmas (e.g., how often certain situations occur, any patterns or trends in the ethical dilemmas occurring)?
- Jot down examples:

WEDNESDAY—TALK ABOUT IT

- Talk to a colleague, supervisor, or mentor about this. You might start the conversation like this: "Can you tell me about a time that you've had to address an ethical dilemma with someone?" or "What do you do if you've identified that you may have made a decision and potentially engaged in misconduct?" Ask them what steps they take to address the misconduct of others and how they prepare for those conversations.
- You might also consider asking your colleague, supervisor, or mentor if they have had to report the potential misconduct of others or their own potential misconduct to the relevant bodies (e.g., organization, licensure board, certification body). If they are able/willing to share, how did they approach and address the situation? What was the outcome? What would they have done differently?

- Talk to your supervisees and trainees about their comfort level with addressing misconduct by others. Have they had instances when they have had to address it? Have they discussed it in their coursework or supervision? Are they aware of the requirements for reporting misconduct by others to the relevant bodies (e.g., organization, licensure board, certification body)?
- Talk to your supervisees and trainees about how they address their own potential misconduct. Are they aware of the requirements for reporting their own potential misconduct to the relevant bodies (e.g., organization, licensure board, certification body)?
- Jot down notes and thoughts from your conversations:

THURSDAY—ACT ON IT

- On Monday you reflected on (or wrote down) some reasons why you might not have addressed the possible misconduct of others. Did you identify any areas you could work on? For example, is it practicing having these difficult conversations? Or is it taking time to prepare how you might approach the conversation? Make a list of some actions you'd like to take (e.g., practicing conversations with your mentor, writing scripts for approaching these conversations), and plan out some time to work on these items.
- You also took some time to reflect on the outcome of instances when you did address the potential misconduct of others. Did you identify any things that could have gone better or could have been done to make the conversation more successful?

  Make a list of the actions you'd like to take (e.g., role-playing similar conversations, scripting different responses), and schedule time to work on these tasks. You might find that there are some similarities between these items and the ones you listed out when reflecting on instances where you did not address the potential misconduct.
- Did you identify any other steps or skills you'd like to add to your to-do list from the discussions on Wednesday about addressing

the potential misconduct of others? How about addressing your own potential misconduct?

- You might consider creating a list of ways you can continue to check in with yourself to identify any potential misconduct and hold yourself accountable.
- You might also consider spending some time making sure you are aware of all the relevant requirements for reporting potential misconduct and schedule a standing event (maybe every 6 months) to review requirements to make sure you are up to date on any potential changes.
- How do you plan to teach your supervisees and trainees to address these situations? Take some time to list out some ways you can teach them how to respond in these situations.
- Jot down thoughts and tasks:

FRIDAY—REFLECT ON IT

- This has been a heavy topic! It's not easy to think about situations when you might have to address potential misconduct either with yourself or others. Take some time to give yourself some room to identify how you feel about having spent a week working through this topic. Has it alleviated any of your stress around this topic? Has it made it worse? If it has made it worse, take some time to reflect on why that might be the case.
- Do you feel like you have a better understanding of how to address potential misconduct in others? How about your own potential misconduct?
- Do you feel like you're more aware of the requirements to report your potential misconduct to the relevant bodies (e.g., organization, licensure board, certification body)?
- How will you continue to hold yourself accountable in these situations in the future?
- Jot down something new you learned or something that surprised you:

Resources to Check Out

- There are several books that can help you build the skills needed to have challenging conversations like the ones we've been talking about. Here are three to get you started (some may look familiar):
  - *Crucial Conversations* (Grenny et al., 2022)
  - *Difficult Conversations* (Stone et al., 2023)
  - *Radical Candor* (Scott, 2019)
- As we mentioned above, the *Ethics Code for Behavior Analysts* (BACB, 2020a) provides a model for addressing ethical dilemmas.
- In their article, Rosenberg and Schwartz (2019) propose a six-step ethical decision-making model.
- Kelly, Greeny, et al. (2023) conducted a study to learn how behavior analysts navigate ethical dilemmas, including what resources they use. You can use the information provided to compare against your own behavior or the approach commonly used at your organization.
- Bailey & Burch (2022) cover ethical decision-making and provide scenarios and case studies for the reader to familiarize themself with the process of addressing ethical dilemmas.
- The ethics textbook by LeBlanc and Karsten (2024) provides instruction on practical ethical decision-making for behavior analysts in training who will provide ABA-based services and/or conduct related research. The content covers the history of ethics, then walks the reader through each of the sections of the *Ethics Code for Behavior Analysts* (BACB, 2020a), focusing on applied decision-making.
- Kelly, Shraga, and Bollinger's (2023) textbook guides readers through interpreting and applying ethics standards to common ethical dilemmas.

Week 20

Check in—Last week might have been a lot to take in. You spent a lot of time delving into reflecting on past situations, conversations, and maybe some uncomfortable interactions and outcomes. As you've worked through the last week, you might have identified that there have been times that you should have addressed potential misconduct (in either yourself or others) and did not. Or maybe you realized that you could have addressed it differently. But we bet as you have reflected on those instances and identified some actionable steps you can take, you've set yourself up to address these situations in the future more successfully. Maybe you've even identified some instances when you have addressed these situations successfully.

So, take a deep breath, because as we mentioned at the start, you're going to continue to come across ethical dilemmas and potential misconduct throughout your career. Use this opportunity to start preparing yourself to successfully navigate those situations. Now on to the scenario!

Scenario—Read through the scenario and work through the prompts we've provided. As a reminder, the goal is not to identify the specific standard from an ethics code that was violated but to use the skills you've been developing to think critically about *any* situation and see the ethics-related content and actions.

> *Kai has been working with their client Eunice for 3 years. They have a great rapport with Eunice's family, and she has made a lot of progress. Eunice's family has been on board with all the programs in place and has been open to all the programming changes made to date, with very few questions or concerns.*
>
> *Eunice has started demonstrating an increase in severe destructive behavior, frequently breaking furniture, toys, and learning materials. It has become more and more difficult for sessions to be conducted in the same room as other clients. Kai is worried that the increase in severe destructive behavior is going to impact the progress Eunice has made and wants to update the behavior-reduction protocol as quickly as possible.*
>
> *They do not have a caregiver meeting scheduled for another month and decided to go ahead and conduct a functional analysis before the next meeting and fill the caregivers in when they review the results at the next meeting. The functional analysis identifies that the*

function of the severe destructive behavior is primarily attention, and they write up the results of the assessment of the proposed protocol for addressing the attention-maintained destructive behavior. The meeting went well, and the caregivers provided consent for implementing the new behavior-reduction protocol.

Two weeks later, Kai attended training on obtaining informed consent and realized during the presentation that they made a mistake by not getting informed consent from Eunice's family before conducting the functional analysis, even though they had not objected to any assessments or interventions in the past. Luckily, the review of the results went well, and Eunice's caregivers didn't ask any questions. Kai is concerned that if they tell their supervisor, they'll be asked to report to the state licensure board and certification body. Kai isn't sure what the reporting requirements are for either of those entities. Because the caregivers didn't raise any concerns, Kai decided to make a plan for getting consent in the future but decided not to take any further action as the risk of harm is low, based on their assessment.

- **VALUES CHECK**—If you had to take a guess about Kai's values, what might one or two of them be, and how can you tell based on the scenario? Do they align with your values?
- **IDENTIFY RISKS**—Circle language indicating a risk was present, and then write down the risk/resulting harms.

Did Kai identify potential risks? Did they take action to minimize or avoid the risk? Are there other risks or actions you can think of that could be related to this situation? Was Kai's assessment of a low risk of harm because Eunice's caregivers did not raise any issues accurate?

Be sure to consider specific aspects of the scope of risks or harms!

- ***How big/bad is the risk/harm?***
- ***Who is/would be impacted?***
- ***Was the risk/harm possible, or did it actually occur?***
- ***Is it likely that the actual or possible risk/harm will occur again in the future?***

- **FIND ANTECEDENT STRATEGIES**—Go back through the scenario, and draw an arrow to indicate anything you think functioned as an antecedent strategy. Make a list of the antecedent strategies, and evaluate if they were effective at preventing things like risks or actual harms. How do you know? Could anything have been improved or added?
- **IDENTIFY RED FLAGS**—Underline any red flags that you find, even if Kai did not notice them. If they did not notice them, write down what factors may have contributed to their ethical blindness (e.g., time constraints, lack of knowledge, authority, common practice, perceiving that there was no risk, lack of oversight). Can you think of any others?
- **FIND INSTANCES OF SOMEONE ADDRESSING THE ISSUE**—Put an * next to any language indicating that Kai took specific action to address the issue. Make a list of those actions and then describe if each was effective or not, how you know, and anything you would have changed or added. Were there other points in the scenario where you think Kai could have taken a specific action to address the situation? If so, what would that look like, and how would you know if it was successful? How do you think the meeting with the supervisor might go if Kai shared their misstep? Do you think this rises to the level of needing to be reported to any other relevant bodies (e.g., licensure board, certification body)? How could Kai find out if it rises to that level?
- **CONNECT**—Imagine how the meeting might have gone with Kai's supervisor. How might have the supervisor responded when Kai shared the information? Are there any steps that Kai could have taken to address the situation before meeting with their supervisor? What if the potential misconduct (i.e., not obtaining informed consent) had been identified by a peer or colleague, and they attempted to address it with Kai? How do you think Kai might have responded?

Taking Care of Business—Take some time to review the actions you identified last week. Do you have time scheduled this week to work on them? How will you continue to work on this in the future? You might calendar some standing time now, so you have protected time to continue to reflect on and work on these items. Also, make sure to teach your supervisees and trainees how to work through these situations too!

Words of Inspiration—You are here and engaging in daily ethics practice because you care about the people you serve and the quality of the services and supervision you provide. And look, the longer you practice, the more likely it is that you will have to address someone's misconduct, or your own. You might find moments when you are overwhelmed, but you are doing the work, and you'll find that it will get easier over time.

Week 21 & 22—Focusing on Compassion in the Therapeutic Caregiver Relationship

Week 21

Check in—Hey there, how are you doing? You just spent time in heavy reflection about the possible misconduct of yourself and others. That is big work and takes a tremendous amount of honesty and courage. You're doing great! We're going to shift direction a bit to revisit our engagement and responsibility in developing effective therapeutic rapport. We are excited that you are here and sending you a big high five. Here we go!

Focus for the Week—In Weeks 5 & 6, we talked about the importance of caregiver collaboration and establishing a strong therapeutic relationship from the onset of services. Caregivers are the primary support network and can drive change in partnership with us as service providers. During the initial intake and assessment process, we have the opportunity to establish that relationship, but starting strong is just the beginning.

How do we maintain, manage, and repair therapeutic rapport as we partner with caregivers and families? What is important to have in place, watch for, and support with? As we implement programming, collect data, and analyze changes and improvements in partnership with caregivers, it is equally important to engage in the soft skills that help us manage and maintain a healthy, collaborative relationship with families. At the time of writing, more than half of behavior analysts certified by the BACB have received their certification within the last 5 years (BACB, n.d.), and there is limited formal training for behavior analysts on the importance of effective therapeutic rapport and compassionate care (LeBlanc, Taylor, & Marchese, 2020). For these reasons, we look forward to spending more time with you on the importance of involving caregivers and equipping you with tools to facilitate a healthy and strong partnership throughout the duration of service delivery.

Remember way back to Weeks 5 & 6 when we said it's not just what we do, but how we do it? Our technology and expertise as behavior analysts is irrelevant if we don't deliver it in the context of compassionate care and a healthy partnership. Across other healthcare industries, interpersonal skills such as empathy and compassion have been positively linked with social validity and satisfaction, adherence to treatment, enhanced engagement, and improved clinical outcomes (LeBlanc, Taylor, & Marchese, 2020).

Taylor and colleagues (2019) described compassion as action-oriented empathy, which harnesses our awareness of another's experience as it relates to our own and translates it into action. For example, we can engage in empathy when we become aware that a caregiver is discouraged because their child is not vocally communicating. We can call up our own past painful experiences with feeling discouraged. With compassion, we show up to actively listen, problem-solve, and collaboratively support next steps and outcomes.

Compassion needs to be the common thread through our communication, demeanor, planning, considerations, thoughts, and priorities. We should see through a lens of compassion as we seek to understand and serve others in the way that they need. As we navigate and coordinate compassionate care, part of managing that rapport is proactive communication, while also monitoring for any risks or red flags that could threaten the established rapport and impact access to care.

Penney et al. (2023) propose that compassion should be the eighth current and essential dimension of behavior analysis. They also identify five guiding principles to be considered in our delivery of compassionate care and implications for practice, which we suggest you utilize as a resource. The first is beneficence, which is our focus to maximize outcomes for clients in a way that is socially significant, appropriate, and important (Wolf, 1978). They also reiterate the importance of staying focused on meaningful outcomes that incorporate clients in the decision-making process, which is our true goal. It is so easy to get stuck on a protocol, procedure, or graph, but if the applied benefit is not realized, we are missing our purpose entirely.

Another important guiding principle is inclusion, meaning how we should think about where ABA belongs within our client's current life, as opposed to fitting a child into an ABA model (Penney et al., 2023). Penney and colleagues (2023) define professional excellence as going beyond practicing within our scope of competence and practicing with a sense of humility and compassion. We have big requests of families. ABA-based services are not always easy to implement, but they become easier in an established relationship that is built on trust, and we "make friends before we make changes" (Penney et al., 2023, p. 8).

Self-determination as a guiding principle focuses us in on the importance of using caregiver and client priorities to inform treatment and focus on areas that they believe are important. Dr. Linda LeBlanc often says that we should almost always compromise and support caregiver input, because it is not our kid who we are programming for (LeBlanc, Sellers, & Ala'i, 2020).

Values Check—Behavior analys is is based in values and the focus to use our science to improve the quality of life for the clients who we serve. Reflect on this foundational value of our science and how it aligns with your own. How does this show up in your life, and more specifically, in how you show up? Do you believe that a strong therapeutic relationship creates a foundation and infrastructure to build from and support better outcomes?

What are your thoughts around family input and your role in treatment? Is the goal to collaborate and compromise within assessment and treatment, or to dictate the recommendations and plan to the family? Do you believe that your role is both learner and teacher? What role do empathy and compassion play in an effective therapeutic relationship, and how are your values aligned with this? Do you value compassion and show it not only to others, but yourself as well? How do your values around caregiver engagement influence the level of priority you place on the ongoing management, engagement, and monitoring of the relationship? These are heavy questions to ask yourself, but they will also highlight areas that you are more likely to lean into and away from and the related impact that could have on those you interact with.

Risks & Benefits—The services that we provide are relationship based, which means that we need to show up every day in a way that supports effective therapeutic rapport. Healthy caregiver relationships are the glue that hold service delivery together. There is tremendous risk when we ineffectively manage caregiver rapport and collaboration for ongoing services. The risk of ineffective care coordination often results in disengagement, including inconsistent attendance, lack of participation, inconsistency in treatment integrity, or withdrawal from services completely.

On the other hand, the benefits of the effective ongoing management of caregiver engagement are endless. Active ongoing caregiver involvement allows us to ensure that our efforts are directed in a way that is well matched to, and driven by, goals that are meaningful to the natural context. In this way, we can live our ethical obligation to place the client first and to individualize programming in ways that matter to the client and their caregivers.

Antecedent Strategies—The key to avoiding most downstream problems and potential risk is to take a proactive approach. This includes anticipating possible needs or concerns, and communicating clearly, consistently, and compassionately. It is critical to establish a relationship based on trust so that caregivers can voice their concerns with us as soon as they come up. As behavior analysts, we know the power of using effective antecedent strategies to increase the likelihood of families (and ourselves!) contacting reinforcement from the beginning and to set the stage for success throughout the duration of services. One strategy is to communicate early and often.

It is always best to regularly check in with families before there is a concern and structure the environment to facilitate open communication. Proactive communication also creates a space of shared knowledge between the family and service provider that supports transparency and management of expectations to minimize the likelihood of discontent down the road. Here are a few things to consider:

- **Think of the most likeable person you know. The person who you want to share good news with, the person who you want to laugh with, the person who you go to for wisdom and advice. What qualities do they hold? Do you demonstrate them when you interact with caregivers?**
- **Are you present during arrival to say hello, check in with caregivers, and ask how things are going?**
- **Do you intersperse more formal check-ins and progress reviews with check-ins on how life is going?**

- **When was the last caregiver meeting, and what were signs of positive engagement and success to suggest the likelihood of continued engagement? In other words, why would they continue to engage with you?**
- **When was the last time you shared a success related to progress with programming?**
- **What was the last client picture or video that you shared with the family that highlighted a newly acquired skill or decrease in challenging behavior?**
- **When was the last time a caregiver gave you feedback or disagreed with a goal or intervention strategy?**
- **How do you match the modality of your communication with the message that you are sharing? For example, would you text a caregiver to communicate updates to a new program and explain the rationale, or is that more appropriate for a recurring, in-person meeting? Have you asked families about their preference for communication modality?**

We've talked about the importance of how we show up, including what we say, what we do, and what we don't do. These interpersonal skills and strategies are critical to establishing and maintaining effective rapport and promoting caregiver engagement. These antecedent strategies and others mentioned in the "Resources to Check Out" section are tools that are available to use, to minimize the likelihood of miscommunication and potential damage to the ongoing relationship. Despite our best efforts, miscommunication does happen. Whether we called the wrong number to cancel a session, or the team member did not communicate there were no remaining diapers, or something more significant in which a program was modified and not clearly communicated to the family.

It is not how we avoid conflict but how we respond and move through it. When miscommunication occurs, lean in, validate, own, fix it, and move forward.

The same principle of communicating early and often applies here. The power of investing in the early foundation of the relationship is an established strong foundation that can withstand the winds of conflict and miscommunication. We cannot completely remove the possibility for miscommunication, error, or conflict, because we are humans and humans are complicated. Instead, our goal is to support continual caregiver involvement so that we can swiftly identify, address, and communicate a resolution and plan. It is not how we avoid conflict but how we respond and move through it. When miscommunication occurs, lean in, validate, own, fix it, and move forward.

Red Flags—As we discussed in Weeks 5 & 6, the initial rapport building at the beginning of services is a critical opportunity to establish expectations, rapport, and a relationship that facilitates ongoing coordination and engagement. If that hasn't happened, that is your first red flag, and we encourage you to review that earlier biweekly pair. There can be so many potential red flags in this area that we are going to categorize them into two buckets: red flags that we observe in ourselves, and red flags that we observe in, or from, others. It is easy to focus on others and point fingers. You set up a parent meeting and they canceled. You asked them to collect data and they didn't. So, let's start with red flags that we create by what we do or do not do.

Remember how we said that self-reflection is a superpower? As you think about it, consider how often you reflect on a conversation or interaction with another person, and identify what you could have done better. How did you contribute positively to the interaction, and how would you engage differently in future interactions? How often do you reflect on the impact that you have on others, and what you can do better? If your score is on the lower end, that's okay because you're honestly reflecting on your ability to self-reflect, and you know this is a red flag for you. If your self-reflection is low, you will likely be less aware of the impact that your behavior has on others, which influences the quality of the relationship and the likelihood of continued interactions.

Other red flags that may exist are things that YOU do that could be negatively perceived or become a barrier to effective therapeutic rapport. This could include miscommunication or no communication at all, being late, saying that you will do something and you don't, having minimal or flat affect when you interact and engage with caregivers, lying, not engaging in active listening skills or applying any feedback provided by the caregiver, talking too much during meetings or not at all, seeming distracted and multi-tasking, or using technical language that is difficult to understand. How do you feel when a professional engages in those behaviors during interactions with you? Does it make you feel valued, supported, and heard? Use your self-reflection superpowers—what opportunities do you have?

Okay, so enough about us. Let's look at the red flags around us that are found in the behaviors of others. These could indicate an opportunity to repair rapport that may be damaged, in a compassionate and supportive way. To effectively manage and maintain caregiver relationships, keep an eye out for the red flags below:

- **Lack of engagement and commitment** are red flags that may signal there is a problem occurring and an opportunity for improvement. This may include frequent cancellations to scheduled appointments, lateness or picking up early, cancelling or frequently moving caregiver meetings, and lack of engagement and consistency when implementing strategies that have been reviewed to support generalization and maintenance of skills.
- **Change in engagement** is unique to each caregiver. If there is a caregiver who has consistently been vocal, excited, and interactive and taken initiative to share successes and ask questions, but the quality and frequency of communication immediately decreases, that is probably a red flag. If there is a caregiver who has been consistently quiet, delayed in responding, but vocal about their gratitude who in contrast begins asking repeated and detailed questions about scheduling practices, that is a red flag.

  The key is the demeanor, communication, frequency, and interactions with the caregiver have changed. That red flag indicates an opportunity to reach out and check in, which may indicate no concerns, or an opportunity to meet the question and support the resolution as a team. This may sound very intuitive reading that here, but when there are competing contingencies of life and work, this red flag can be easily missed.
- **Grievance, conflict, and concerns** will happen, and again, it is less about avoiding them completely and more about how we respond to them. If there is a caregiver complaint about team performance, pro-

gramming, progress, attendance, or any other component of service delivery, this is an opportunity to contact the caregiver to evaluate the concern and intervene, while using your crucial conversations skills. These issues create roadblocks that can either get us off course or delay access to care. Our goal is to address and resolve them as quickly as possible, then refocus to manage and/or repair the relationship.

- **Surprises (not the good kind)** were discussed in the initial stage of caregiver engagement but are equally important in the ongoing management of care coordination. Our ethical responsibility is to integrate and inform caregivers on all aspects of their child's programming and experience. Red flags are any statement or action that indicates that is not the case. This may include lack of knowledge as to what their child is working on and the focus of treatment, a change to their behavior intervention plan, or specific strategies implemented that are not being implemented at home and in other environments. Other examples include a change to the recommended and provided intensity of care outlined in the treatment plan that's not discussed, or an updated treatment plan that included no input from the caregiver in the selection of goals and treatment scope.

Addressing—If you or your workplace is engaging in less-than-optimal behaviors to support effective ongoing therapeutic relationships that prioritize engagement and collaboration, we encourage you to systematically and thoughtfully review opportunities that exist to improve your daily practice. Develop a list of quick wins and items that may take more time and coordination across multiple individuals. You might determine a quick win is to ensure that ongoing caregiver meetings are scheduled, and that you have weekly check-ins to celebrate progress with caregivers. This week you might review your caseload, check in, schedule, and share an updated agenda with families that highlights the plan, progress, and space for input to evolve and align progress that is supported from the expanded network actively supporting the individual who we serve. Maybe you'll spend time evaluating the status of caregiver engagement and therapeutic rapport to triage and identify action items to manage the relationship or potentially repair any damage that is identified.

We also encourage you to review the "Resources to Check Out" section below on compassionate care and evaluate the extent to which you integrate this approach into your practice. As you think about how you'll address identified opportunities, also consider how you will match the message to the modality. If you have identified damaged rapport, you will likely want to speak in person or on the phone with the caregiver, as opposed to an email. If you are sharing a minor update to the schedule, an email is probably fine. If there are bigger and broader opportunities to address, we encourage you to break them down into manageable next steps and take the first one. If your organization does not have a policy or process to manage consistent cancellations, you will likely communicate this to your supervisor and/or other clinical leaders and potentially volunteer to be a part of the workgroup that will create this structure and resources for the team.

Weekly Actions

MONDAY—THINK ABOUT IT

- Reflect on your current relationships and rapport with caregivers. How often do you communicate with them? What strategies do you use to gather their input about things like assessment procedures, goals, targets, and intervention strategies? How often do you engage in formal collaboration and solicitation of their input and opinions?
- How does your favorite person engage with you? What qualities do they possess, and how is that similar or different to how you engage with others?
- How do you live your values and make decisions in which you show up as the person and professional that you want to be?
- Where can compassion be seen in your life, in how you engage with both others and yourself?
- What do you think about the guiding principles shared and the proposed eighth current and relevant dimension of behavior analysis (Penney et al., 2023)?
- Jot down notes and thoughts:

TUESDAY—LOOK FOR IT

- Observe other professionals engaging with families during session pick-up and drop-off. What do they do very well? What missed opportunities do you observe to influence the interaction?
- Review your calendar from the last 3 months. Did you plan for caregiver engagement? Did the meetings and support that you had planned actually occur?
- How does your organization collect caregiver satisfaction data? How often? How are those data used?
- Are families engaged in ongoing treatment and progress? Do they show up to meetings with you and for scheduled services? If they do, that's a green flag! If not, do you see red flags?
- When was the last time each caregiver you partner with shared a success they were excited about?
- Jot down notes and thoughts:

WEDNESDAY—TALK ABOUT IT

- Find a colleague or group to reflect with on what you have observed and found.
- Talk with your supervisor about the current expectations and resources available to effectively manage and maintain therapeutic rapport. Ask them how they think about compassionate care and the role that it plays in ABA and our service delivery. Share ideas with them about your self-reflections.
- What ideas and actions have others taken to facilitate an effective ongoing therapeutic relationship? What red flags have others observed in themselves and within the relationships they support?
- Talk about compassion and compassionate care with other behavior analysts. Share the article with them from Penney et al. (2023), the evolution of our field, and the importance of compassion being at the center of our practice and engagement with caregivers.

- Share an example of excellent therapeutic rapport and compassionate care that you have supported. What were the benefits, both short and long term?
- Jot down notes and thoughts:

THURSDAY—ACT ON IT

- Think about the red flags that you identified within yourself and how you will minimize the likelihood of the red flags in the future. Or use the list to create a checklist or rating scale for you and your team.
- Now, think about the red flags that you identified based on your reflection and evaluation of the current therapeutic relationships that you have. Is there a lack in engagement or commitment to care? Has there been a change in engagement or communication? How will you continually monitor and evaluate this?
- What training and resources can you share with your supervisor and/or the organization that you work for?
- Create a caregiver survey to ensure that you are capturing the guiding principles of compassion in the way that you would hope and want others to experience. Are they active participants in their child's care? Do they value the partnership they have with you?
- Jot down notes and thoughts:

FRIDAY—REFLECT ON IT

- How did your review and analysis inform your action items and plan of action?
- Relationships are complex and each is different, which can easily become overwhelming. It may be helpful to color code or triage the easy wins and then plan out pieces of the larger initiatives.
- Therapeutic rapport is built on trust, consistency, and communication. How have you managed trust in your caregiver relationships? How have you worked to repair broken trust or damaged relationships?

- Jot down something new you learned or something that surprised you:

Resources to Check Out

- Dale Carnegie (2022) does a stellar job in his classic book *How to Win Friends and Influence People*, providing discrete tips and tricks to consider in how you show up with people. We included this resource in Weeks 5 & 6, but it is also valuable in the ongoing relationship.
- Linda LeBlanc, Bridget Taylor, and Nancy Marchese (2020) do a beautiful job highlighting the importance of compassion in our work as behavior analysts. We recommend specifically using the tables embedded in the resource that could even be translated into checklists.
- Spend time with the oldie-but-goodie article by Wolf (1978). He was shouting about compassion way back in the '70s, and you'll likely find his words relevant today!
- Consider compassion as the eighth dimension of behavior analysis with Penney et al. (2023). This resource will provide further color and ideas to incorporate a compassionate approach into your ongoing therapeutic rapport.
- Again, *Crucial Conversations* (Grenny et al., 2022) is a great resource to support your ability to have difficult conversations with caregivers. Especially if you identify that you need to rebuild rapport or just address a topic that you know has been difficult to address in the past with a caregiver.
- Remember those soft skills you reviewed in the Taylor et al. (2019) article? Well now might be a good time to give yourself a little refresher—those skills are relevant throughout the entire caregiver relationship and might require some practice to maintain, especially when conversations get difficult.
- Rohrer and colleagues' (2021) resource can be used throughout services, not just at the start of the relationship. Use it to evaluate how you're doing with maintaining a compassionate and collaborative relationship.

Week 22

Check in—That was a big week with lots of content and opportunity to self-reflect, plan, and act. Or, you may still be processing and reflecting, which is great! That is the continual process of lifelong learning—reflect > act > observe > change—rinse and repeat.

reflect > act > observe > change

We applaud you for being here and doing the difficult things, and showing up in a way that is better than yesterday and in a way that commits to ongoing, purposeful caregiver involvement. The soft skills that we talked about (e.g., interpersonal skills, effective communication, warmth, kindness, and compassion) can sometimes be the hard skills. Let's read a bit more in the following scenario.

Scenario—Read the scenario and complete the prompts. Our focus is not to identify the specific standard from an ethics code that was violated, although that is a great activity! The purpose is to think critically about *any* situation and see the ethics-related content and actions. How can you relate, how can we reflect on similar opportunities, and how we will mitigate risk?

> *Ben is a certified behavior analyst providing care to a 2-year-old client, Leo. Leo's mom's name is Persian. Ben and Persian met to complete Leo's initial assessment, and he has been receiving services for 3 weeks now. A caregiver meeting has not been scheduled since the initial treatment plan review.*
>
> *Persian has asked to meet with Ben for the last 3 days when she has dropped Leo off at the center, but Ben has not been available. She gets ahold of Ben the following week via phone and communicates that she has not heard from Ben. Ben indicates that he has been really busy, because his caseload is too big. She asks to set up a meeting, and he says sure.*
>
> *The following Monday, Persian brings Leo to the center at 8 a.m. for his session and to meet with Ben. It looks like Ben is running late, so Persian spends some time with Leo and his behavior technician.*

She notices that Leo is working on programs that she is not familiar with, and the targets in his following directions program are things that he readily does at home. Ben shows up at 8:14 and apologizes for being late—there was lots of traffic.

Ben and Persian sit down, and Ben jumps into talking about some of the challenging behavior that Leo has been demonstrating and what they are doing to "figure it out." He talks for another 20 minutes before Persian stops him and asks if she can ask a few questions and provide some feedback. She indicates that when they initially spoke, Ben reassured her of the importance of her feedback, engagement, and partnership in services delivery. She tells him that she's disappointed to see that he is working on programs that she is not aware of, and the programs that she is aware of are far too easy for him. There seems to be a disconnect and a missed opportunity to collaborate in a way that is best for Leo.

Ben interrupts her and says that he understands and will make an effort to communicate better. He then asks Persian if she has any thoughts or feedback on areas and programs that could be better supported. She shares that she would love to begin talking about toilet training, to improve his independence and get him out of diapers, which are not helping a rash that she and the doctor are trying to get rid of. Ben says that he doesn't think it is time and Leo is too young. Persian leaves the meeting feeling very disappointed and confused. She does not feel heard or respected and does not trust that Ben has Leo's best interests in mind. She cancels several of Ben's sessions over the next 2 weeks and ultimately chooses to withdraw him from care, as she does not feel like a valued partner in Leo's care.

- **VALUES CHECK**—Do you think that Ben took the opportunity to identify his values, or act on them? What does he appear to value, and what opportunities does he have as a professional and behavior analyst? Have you interacted with BCBAs who have similar values or behavior to Ben?

- **IDENTIFY RISKS**—Circle language indicating a risk was present, and then write down the risk/resulting harms.

  Did Ben identify potential risks? Did he take action to minimize or avoid the risk? Are there other risks or actions you can think of that could be related to this situation?

Be sure to consider specific aspects of the scope of risks or harms!

- **FIND ANTECEDENT STRATEGIES**—Go back through the scenario and draw an arrow to indicate anything you think functioned as an antecedent strategy or opportunities where an antecedent strategy was missed. There were many missed opportunities in this scenario. List them out and determine what else could have been improved or added.
- **IDENTIFY RED FLAGS**—Underline any red flags that you find, even if Ben did not notice them. What factors may have contributed to his ethical blindness (e.g., time constraints, lack of knowledge, authority, common practice, perceiving that there was no risk, lack of oversight)? Can you think of any others? Persian did a nice job communicating her concerns and feedback, but what red flags and consequences would have existed if she did not do that?
- **IMAGINE INSTANCES OF SOMEONE ADDRESSING THE ISSUE**—Put an * next to any language or missed opportunities in which Ben could have responded differently. Imagine if he had responded differently in a way that was compassionate, warm, and established trust. How would caregiver engagement have been different? How would outcomes for Leo have been different?
- **CONNECT**—Reread the scenario and think about the role and responsibility that Ben played in this ethical violation. Reflect on similar situations that you have observed or been a part of and how you will structure your environment to integrate these learnings into your practice to mitigate risk for yourself and minimize vulnerability to those you serve. How will you own your ethical responsibility in compassionate care and caregiver collaboration to support the best possible client outcomes?

Taking Care of Business—Think about your reflection and what is similar or different to what you are doing today. Work with your team or colleagues to ensure that interactions with caregivers embrace the importance of soft skills, patient care, and overall experience. How do policies and programs set both families and team members up to be successful; to clearly define and outline what caregiver engagement is, what it looks like, and what is expected from all parties; and to explain the outcomes of not doing so? Review the aspects of your current rapport with caregivers, and evaluate what you are doing very well and what can be improved. Reflect on the level at which you show compassion to yourself and those around you. Take stock of how often you ask caregivers for direct feedback, opinions, and input. How does your empathy and ability to relate to others prompt action to make their lives better?

Words of Inspiration—Compassion is kindness with wheels and a vehicle to bring others to a better place. Travel well.

Weeks 23 & 24—Connecting With Culturally Responsive Practices

Week 23

Check in—How are you doing? We started out this set of biweekly pairs by discussing addressing ethical dilemmas with others and yourself and then jumped right into caregiver collaboration. Those can both be heavy topics, especially if this is your first time incorporating thinking about these topics into your daily practice. So, take a moment and check in with yourself. Are you ready to continue? Or do you need to take a brief pause before you do? If you need to pause, you could take a day or two before starting, or even jump ahead to the rest and reconnect weeks coming up next and then come back to this topic.

Focus for the Week—Before we get started, we would like to acknowledge that we are not experts in this topic. All three of us are still in the process of learning and acknowledging how we can continue to improve our knowledge and skills to provide culturally responsive supervision and services. We are on this journey with you.

To help you prepare for this biweekly pair, you may want to refer to the Chapter 3 section "Prepping for Implicit Biases and Cultural Responsiveness" and Weeks 15 & 16, which covered personal biases. This may be a topic that might result in some discomfort, especially while you engage in some self-reflection. Make sure to approach this topic with self-compassion and the goal of doing better in the future. You can't change your past behavior. Yes, it is important to take time to self-reflect and identify areas where we can continue to grow and improve. That said, there is little benefit in getting stuck in those past moments. Acknowledge them and use them as motivation to continue your journey.

Not only is it important that we acknowledge and address our past experiences and how they have shaped us, but it is also important that we acknowledge that everyone comes with their own set of experiences, perspectives, and values that likely differ from our own. Especially when thinking about providing supervision and services, we must make sure we are culturally responsive in these relationships. While working through the biweekly pair on implicit biases, you likely identified some of your implicit biases. And you also spent some time thinking about how to address these biases, right? In doing that, you've already taken an important step in your journey to cultivating cultural responsiveness.

And remember what we talked about in Chapter 3? You might be at the beginning of this journey, or maybe you've been working on this for a while. Also, you might identify that there are some areas where you are further along in your journey than others (e.g., maybe you've really focused on this with your clients but are at the start of this journey in your supervision practices). The goal is to continually assess where you are and then use that information to inform your next steps (e.g., determine if you are practicing outside of your scope of competence; identify if you need to reach out to a colleague, mentor, or supervisor for support).

The American Speech-Language-Hearing Association (n.d.) provides the basis of definitions for three important terms for us in this section:

Cultural competence refers to the process of engaging in self-assessment and education to learn about others' values and beliefs and sharing your own values and beliefs.

Cultural humility refers to engaging with and exploring your own cultural identities, values, and beliefs to better understand the cultural identities, values, and beliefs of others.

Cultural responsiveness refers to acknowledging, understanding, and responding to the unique and intersecting cultural variables that each person has.

What we want to highlight here is the common theme that engaging in all three of the above is an ongoing process and will involve self-reflection and active engagement in this area. Just because you talk to a client and the client's family or your supervisee once about their perspective, values, and experience does not mean that you are done. Humans change constantly, and so do their perspectives, values, and experiences. This is why it must be an ongoing activity. And you know what? You're not always going to get it right.

So now is the time to practice leaning into your discomfort. Because sometimes you are going to say or do the wrong thing. You'll make a decision that doesn't take the client and caregiver's perspective into consideration. Or maybe your supervisee shares the feedback that you made a comment that made them feel uncomfortable or that their perspective was not valuable or valid. And remember—it's not about the intent behind what you say or do, it is the impact. Don't get us wrong. Intentionally behaving in a way that is harmful is a problem and must be addressed.

The point we are making is that if we are to behave ethically, we must attend to our behavior even when it was not intended. And that can be hard! You're probably not actively trying to be unaccepting, harsh, or inappropriate, and it can be easy to hide behind saying, "Well, that is not what I meant or intended." But it is the impact your behavior has on the other person that is important to acknowledge and address, especially when it is negative. Be kind to yourself as you work through this material. You might identify times that you did not engage in culturally responsive behavior. Take that moment and learn from it, but don't dwell on it; you can't change the past. What you **can** do is take what you learn in that moment to set yourself up to do better in the future.

Values Check—What values did you identify on your list that seem to be a good fit with this topic? You might consider checking out the values you selected when reviewing the section on awareness of your personal biases and interfering factors (Weeks 15 & 16). Again, this topic might be uncomfortable to work through. Are there values that you identified that align with staying motivated and working through the discomfort as you work through this biweekly pair? Take a moment and write them down here, so you have them ready to look back on if you find yourself losing motivation or struggling to get through the material.

Risks & Benefits—Take a moment and remember back to Weeks 15 & 16, when we reviewed personal biases. Remember, just because you identify implicit biases, prejudices, or life challenges, it does not mean you are a bad person. It just means you are a person, like the rest of us. And the same is true if you identify that you may not have demonstrated culturally responsive behavior in the past as a clinician or supervisor. Or if you notice that you are just starting to evaluate your cultural responsiveness and implement it into your practice and supervision.

One risk of not engaging in and connecting with culturally responsive practices is that you are likely to provide services and supervision based on your perspective, values, and experience in the world and not the perspectives, values, and experiences of those you are serving or supervising. This will greatly impact your ability to successfully provide services and supervision to those with diverse backgrounds and probably damage or prevent the development of harmonious relationships.

For example, say you determine it is necessary to put in a program to increase a client's time spent sleeping in their own bedroom. However, you learn that the family you are providing services to prefers to co-sleep for the first few years of childhood. Without asking questions and learning more about the family's values and culture, you might put in a program that harms your relationship and rapport with the family and makes them feel as if their culture is not valued in the behavior-analytic services you are providing.

Additionally, it is risky if you are providing services and supervision within your scope of competence if you are not assessing your own cultural humility and competence and your cultural responsiveness when providing services and supervision. Just assessing these once runs the risk of not taking into consideration that our perspectives and beliefs change over time.

One big benefit of engaging in and connecting with culturally responsive practices is that you will be able to provide high-quality services that respect the culture, values, beliefs, and experiences of those you serve (e.g., clients, caregivers, supervisees, trainees) and protect them from harm.

Moving from a place of cultural humility and responsiveness will likely result in healthier, more trusting, therapeutic relationships with clients and caregivers. The same is likely true of your supervisory relationships. And in teaching your supervisees and trainees how to do the same, you will be showing others how to incorporate this learning and practice into their own supervision and service provision.

Antecedent Strategies— It is time to get comfortable and prepare to engage in self-reflection and assessment. The first way you can prepare and set yourself on a path to connect with culturally responsive practices is to assess where you are on this journey and assess across different areas of your professional work. Think about the clients you are currently providing services for or have served in the past. Think about your supervisees. Did you ask them questions about their culture, values, perspectives, and experience? If not, that is okay! Don't beat yourself up about it. As we have mentioned before and will mention again, the point of you doing these exercises and working through this material is to help you incorporate this into your daily practice and identify areas where you can improve. And guess what! There are resources to help you with conducting self-assessments on your practices. Turn to the "Resources to Check Out" section at the end of this week to see a list of some of the resources that are out there.

Another antecedent strategy is to develop the practice of getting curious about others by asking questions. You need to know about the culture, values, perspectives, and experiences of your clients, their caregivers, and your supervisees and trainees to be able to engage in culturally responsive practice. You might start by practicing asking the questions with a colleague, mentor, or supervisor, especially if you haven't done this before or are feeling uncomfortable at the thought of starting these conversations. And there are resources for this too (see the "Resources to Check Out" section).

The third antecedent strategy is to identify a plan for how you are going to incorporate this information into your practice. How will your knowledge gained through asking questions impact the assessments you select, the goals you identify, the interventions you choose, and the way you provide training and feedback? How will you use it to positively impact the interactions you have with clients, caregivers, supervisees, and trainees?

And don't forget to continue your assessment (of yourself and those you serve). As we've already identified, people change, as do their experiences, perspectives, and values. And these changes may impact the services and supervision you provide.

Finally, how are you going to make sure to be kind to yourself? You're not going to get this perfect. There may be times where you receive some uncomfortable feedback because you've missed the mark, engaged in behavior that has made someone uncomfortable, or selected an intervention that did not take into consideration the cultural values of the client and family. And if that happens, you need to find ways to effectively address it AND show yourself kindness. The goal is continuing to learn and improve how you are providing culturally responsive services and supervision.

Red Flags—One giant red flag is if your immediate reaction is *I am already doing an excellent job at this, and I don't need to conduct an evaluation of how I'm doing.* We'd guess that if you're working your way through this book, it is unlikely that you are having that thought. But just in case, know this is a red flag. Remember, we are talking about an ongoing process of self-assessment and assessment of those you are serving and supervising. Culturally responsive practice is a moving target, not something you are going to accomplish once and be done.

Another red flag is if you find yourself thinking that this is going to take too much time and that what you're doing now is just fine. Keep in mind that you may not notice the impact immediately, but missing the opportunity to take into consideration cultural differences and perspectives can impact the quality of services and supervision provided and the success of both. And finally, if you're finding yourself very uncomfortable and avoiding these practices, either because of lack of knowledge and experience, or worry of what you will learn about your past or current practices in the process, that is another red flag to note and address.

Addressing—If you find that you have not been connecting with and engaging in culturally responsive practices, you can change it. How might you identify this? It might be from reading this biweekly pair. Or maybe you identify it as you engage in some self-reflection and self-assessment using the resources below. Or maybe you received some feedback from a client, caregiver, supervisee, or trainee indicating that you were not engaging in culturally responsive practices with them.

The first thing you can do is acknowledge it. You might want to reach out to a mentor or supervisor first to talk through what next steps to take. Maybe you'll even practice the conversation you're planning to have with them. And then it might be appropriate to meet with the individual impacted and acknowledge it directly. Maybe you have a plan to share how you will move forward. Or maybe you are just acknowledging it and sharing that you plan to take steps to change your behavior in the future, but you wanted to address it with them immediately to show you're taking action. And when you have that conversation, keep in mind that no matter the intent you had, you must acknowledge the impact that you had on the individual.

It could be that you have been actively working on developing your skills in this area, and you notice that others in your workplace are not. Well, then it's probably time to lean into a difficult conversation with a leader you trust about this topic. You might say something like, "I know that as an organization we value ethical practices, and we are continually striving to improve. I've been doing a lot of thinking about our overall approach to culturally responsive practices and culturally humility. Would you be willing to have a conversation with me about how we think we all are doing and if there are places for improvement?"

Now, take a deep breath. If you've made it that far in addressing this topic—good job! Now, if you don't have a plan moving forward, you can take the information you gained from the conversation to prepare for what you will do differently in the future. And you might also check in with your mentor or supervisor here. And don't forget to identify how you will continue to self-assess and evaluate your practices after you implement the changes you identified.

Weekly Actions

MONDAY—THINK ABOUT IT

- How are you doing? Are you feeling anxious, uncomfortable, defensive, or overwhelmed? Hey—that is okay. Thanks for taking a moment and being honest with yourself about how you're feeling. Acknowledge where you are at, and take some time to prepare yourself to move through the rest of this material. Do you need to make some additional space for some self-compassion practices? You might plan to add a self-compassion activity or practice at the end of each day while you work through this material, and maybe a check-in to evaluate how you're feeling.

- How often have you engaged with this topic? Is this new to you? If so, same as with the biweekly pair on implicit biases (Weeks 15 & 16), take it slow. You might want to take additional time to work through this material if that is the case.
- How do your values support you in committing to evaluating your implementation of culturally responsive practices? How can you use your values to help you work through this topic, even when it becomes uncomfortable?
- Take a moment to reflect on your supervision and training. What is your learning history related to self-assessing your cultural responsiveness in service delivery and supervision, and addressing it if you find that it needs improvement? Did your supervisors discuss this topic with you?
- Think back on your clinical practice. Has there ever been an instance when you conducted an assessment or implemented an intervention that did not take into consideration the client and caregivers' culture, values, perspective, and experience?
- Think back to your previous supervisor relationships. Has there ever been an instance where you engaged in behavior that was not culturally responsive (i.e., did not take your supervisee's or trainee's culture, values, perspective, and experience into consideration)?
- How about in the supervisory relationships where you were the supervisee? Have you experienced a supervisory relationship where your supervisor did not engage in culturally responsive supervisory practices?
- If someone asked you right now, "Where are you at on your journey to be a culturally responsive supervisor and practitioner?" what would you say? And it's okay if this catches you like a deer in headlights. We aren't even sure how we would answer this question. The idea is not to have a polished answer at the ready. The idea is just to engage with the question, turn it over in your mind, and struggle with what you might say. If you come back in 2 months and ask yourself the same question (and you probably

should!), you'll probably struggle just as much and come up with a different answer. And that is progress!

- Do you talk about this topic with your supervisees and trainees and incorporate it in their supervision and training?
- Jot down notes and thoughts:

TUESDAY—LOOK FOR IT

- Can you identify examples of when others did not engage in culturally responsive practices in their service delivery or supervision? Were there, or are there, any red flags? If so, list them out. What were/are the risks (i.e., how did it negatively impact others)?
- Does your workplace discuss culturally responsive practices? Do team members feel safe enough to bring up this topic in conversation?
- Is there training available on how to engage in culturally responsive practices in service delivery and supervision?
- Is there training available related to areas like diversity, equity, inclusion, and accessibility?
- Jot down examples:

WEDNESDAY—TALK ABOUT IT

- It's time to have a conversation with your colleague, supervisor, or mentor about this topic. Note that it is possible that this conversation might be a little uncomfortable for all parties involved. You might want to prepare the person you will be discussing it with in advance, so they can identify their comfort level with the conversation and prepare as they need to. One way to start is with something like, "I'd love to start a conversation about how you take into consideration cultural differences when providing services and supervision. I want to acknowledge that everyone may be in a different place in their practice with this topic, and I want to share that the purpose of this conversation is just to start a conversation about it—not to judge or compare where we are in the process."

- If you know someone familiar with this topic, consider reaching out to them. If you don't, consider reaching out to a colleague, mentor, or supervisor to see if they do and might be able to connect you. You might consider discussing with this person if the topic makes them uncomfortable, and if it does, how they respond to those feelings. This might also be a great time to share any concerns you have with the topic, including if you don't know how to have a conversation about this topic with others (e.g., colleagues, clients, caregivers, supervisees, trainees).
- Review this topic with your supervisees and trainees. If you've identified that you evaluated your supervisory relationships, you might consider acknowledging that and sharing how you will be assessing them in the future. Discuss the risks involved in not engaging with this topic and how to watch for red flags and address them. Keep in mind that culturally responsive practices might be an unfamiliar and/or uncomfortable topic for your supervisee or trainee, so be sure to also include the pieces of self-compassion as you discuss the topic with them.
- Jot down notes and thoughts from your conversations:

THURSDAY—ACT ON IT

- Think back to your reflection and conversations from earlier this week. Make a list of any indicators that you may be at risk (e.g., you have not evaluated your practices, there have been concerns in the past that an intervention you have implemented has not taken the family's values into consideration).
- Identify the relationships in which you should be ensuring you are engaging in culturally responsive practices (e.g., with clients, supervisees, trainees, supervisors). Identify (or write down) which relationships you have evaluated and how frequently you have evaluated them to ensure you are engaging in culturally responsive practices. If you identify any red flags, make a plan to address them in a timely manner.

- Create a general action plan you can use if you identify any red flags. This might include a list of antecedent strategies to put in place and steps to take to address the situation.
- Review the related policies, training, and resources available in your organization for addressing this topic. Do they align with your ethical obligations? Are there any updates that need to be made? And if so, can you bring it up to your supervisor to make the needed updates?
- Jot down thoughts and tasks:

FRIDAY—REFLECT ON IT

- How are you feeling now that we are closing out the week? Were you uncomfortable? Were the conversations you had earlier in the week uncomfortable for either party?
- Do you feel more confident with how you might address these situations than you did at the beginning of the week? If you identified holes in your supervision and training related to this topic, do you feel like you have taken steps to start addressing them?
- What will you do to continue to evaluate your practices; ensure that you continue to engage in alignment with your values; and address any concerns as quickly as possible to minimize negative impacts on the clients, caregivers, supervisees, and trainees you provide services for and supervise?
- Do a quick check-in—have you been showing yourself self-compassion as you have worked through this week? Remember, these topics are tough, but by working through them, you are working toward protecting your clients, supervisees, and trainees.
- Jot down something new you learned or something that surprised you:

Resources to Check Out

- This list is not exhaustive—there are many resources (e.g., books, articles) out there that can be used. This list just includes a few, and some are making a repeat appearance from the "Prepping for Implicit Biases and Cultural Responsiveness" section of Chapter 3 because they are that good!

- You might start with reviewing your results if you completed Project Implicit's IATs (Project Implicit, 2011). You might even consider taking them again to see if there has been any change in your biases since the last time you took them.
- Terminology:
  - The article by Beaulieu and Jimenez-Gomez (2022) provides a nice description of the relevant terminology and provides some recommendations for self-assessment.
  - The article by Tervalon and Murray-García (1998) provides recommendations for moving to cultural humility in terminology rather than cultural competence.
  - American Speech-Language-Hearing Association (n.d.) also provides definitions of relevant terms.
- Background:
  - Fong et al. (2017) provide some background on the importance of this work and highlight barriers that are faced as well as some recommendations.
- Assessments:
  - American Speech-Language-Hearing Association (n.d.) provides checklists including a *Self-Reflection Check-In, Culturally Responsive Practice Check-In*, and *Policies and Procedures Check-In* that may be useful in your practice.
  - The article by Beaulieu and Jimenez-Gomez (2022) starts the discussion on assessing your own cultural background, reviews the previous research in this area, and provides some ideas for future research.
  - Gatzunis et al. (2022) introduced the *Culturally Responsive Supervision Self-Assessment (CRSS)* tool.
  - Leland and Stockwell (2019) introduced the *TGNC-Affirming Skills Self-Assessment Tool.*
- Supervision & Practice:
  - Mathur and Rodriguez (2022) propose a cultural responsiveness curriculum and provide a checklist to assess competency.

- Wright (2019) discussed how to implement cultural humility into practice, specifically in ABA.
- LeBlanc, Sellers, and Ala'i (2020) discuss culture and supervisory relationships in Chapter 4 of their book *Building and Sustaining Meaningful and Effective Relationships as a Supervisor and Mentor.*
- In their 2022 article, Jimenez-Gomez and Beaulieu discuss research and considerations for culturally responsive assessments and intervention.
- Čolić et al. (2022) provide guidelines for engaging in culturally responsive practice with Black families.
- The book *Multiculturalism and Diversity in Applied Behavior Analysis* (Connors & Capell, 2020) covers a broad range of topics related to addressing multiculturalism and diversity in the profession. The authors primarily focus on providing recommendations for interacting with clients and caregivers but do include several chapters on professional development and supervision.

Week 24

Check in—Did you take a breather? How are you feeling as you move into the second week of the topic? If you are feeling overwhelmed from the past week, you might consider taking a break before you proceed with this week. And maybe take a second to reflect on the values you identified last week. In working through these difficult topics, you are incorporating ethics into your everyday practice, and it may not feel like it now, but it will get easier. Or at least a little more comfortable. Now let's get to that scenario.

Scenario—Take a moment and review the scenario and prompts below. Remember, don't get caught up in deciding if the people in the scenario are ethical or unethical. It's more nuanced than that. Use the scenario to identify the risks, benefits, values, and red flags that can come into play in this or other similar situations. You might even identify some actions that you want to add to your list as you work through the prompts.

Sofia's supervisee, Jordan, has recently brought up that the stimuli Sofia is using for teaching with their clients have primarily included pictures of Caucasian children and adults. The stimuli and targets also only include women and men depicted in traditional binary gendered roles (e.g., doctors are men, nurses are women). Jordan also points out that there are no depictions of nonbinary, gender nonconforming, or trans people. Jordan shared with Sofia that providing more inclusive stimuli would allow for better teaching opportunities and would also provide better representation for the clients and families they serve and the team members providing services.

Sofia thanked Jordan for the feedback but shared that she does not have the time or resources to create or purchase all new stimuli. She said that she thinks she is aware of some versions that are freely available, but it isn't a priority right now, as she has just been assigned a new client and needs to spend her time prepping for the start of services.

In their next supervision meeting, Jordan shared with Sofia that her response made them feel like their suggestion was not a priority, and it also felt belittling of their experience as a nonbinary, biracial person. Sofia was surprised by this feedback, apologized, and shared that it was not her intent to make Jordan feel that way. Sofia followed up with her supervisor and shared that she did not think that her response was belittling, but she was just being honest about her capacity.

- **VALUES CHECK**—What do you think Sofia's values are as a supervisor and clinician around cultural responsiveness? Do they align with yours? What do you think Jordan's values are related to cultural responsiveness in service provision? Do they align with yours?
- **IDENTIFY RISKS**—Circle language indicating a risk that was present, and then write down the risks/resulting harms. Did Sofia and Jordan also identify the same risks? Did they take action to minimize or avoid

Be sure to consider specific aspects of the scope of risks or harms!

the risk? Are there other risks or actions you can think of that could be related to this situation?

- **FIND ANTECEDENT STRATEGIES**—Go back through the scenario and draw an arrow to indicate anything you think functioned as an antecedent strategy. Make a list of the antecedent strategies and who implemented them, and evaluate if they were effective at preventing things like risks or actual harms. How do you know? Where there any antecedent strategies that you think Sofia could have implemented that could have changed the outcome? Could anything else have been improved or added?
- **IDENTIFY RED FLAGS**—Underline any red flags that you find, even if Sofia didn't notice them. If she didn't notice them, write down what factors may have contributed to her ethical blindness (e.g., time constraints, lack of knowledge, authority, common practice, perceiving that there was no risk, lack of oversight). Can you think of any others?
- **FIND INSTANCES OF SOMEONE ADDRESSING THE ISSUE**—Put an * next to any language indicating that Sofia or Jordan took specific action to address the issue. Make a list of those actions, and then describe if they were effective or not, how you know, and anything you would have changed or added. Were there other points in the scenario where you think one of the individuals could have taken a specific action to address the situation? If so, what would that look like, and how would you know if it was successful?
- **CONNECT**—What do you think would have happened if Sofia had reached out to her supervisor for support earlier? What do you think would have happened if Jordan had not shared that the response negatively impacted them? Have you ever been in a similar situation or known someone who was? What was the outcome? What would you do differently, or recommend the other person do differently, now?

Taking Care of Business—How was that scenario review? Are you still feeling uncomfortable? Or have you noticed a shift in how you are feeling? Make note of it, even if there is no change. Use the rest of this week to follow up on any tasks you did not complete last week, and create a plan to finish them (even if it takes more time). And be sure to calendar some time to check in on this topic over time. Remember, you're working to move this into your daily practice as a behavior analyst. Keep up the great work! Ethics can be a tough topic, and you're doing a great job working through this material.

Words of Inspiration—You're not going to get it perfect all the time. No one does. If you aren't making mistakes, then you aren't learning and growing. Your willingness to learn and continue to improve is what will make all the difference. You've taken the first step in expanding your awareness and perspective, you've set the goal to continue to acknowledge that people and their beliefs and perspectives change, and you've committed to take action. And don't forget to show yourself some compassion—we are all learning and are in this together.

Weeks 25, 26, & 27—Resting & Reconnecting

Week 25—Time to Rest & Take Care of Yourself

This is your third rest week, which means that you are almost halfway through the content. That is incredible! As with the past two rest weeks, check in with yourself so that you can make the most of planning this rest week. Remember there is just as much value and power in purposefully pausing as there is in forging ahead.

Monday–Sunday Self-Care & Wellness Intentions

Spend a little time mapping out your self-care and wellness intentions across the week. You can scan the content and your notes from Weeks 9 and 17 to help you plan.

Monday Self-Care & Wellness Intentions

____

____

Tuesday Self-Care & Wellness Intentions

____

____

Wednesday Self-Care & Wellness Intentions

____

____

Thursday Self-Care & Wellness Intentions

____

____

Friday Self-Care & Wellness Intentions

Saturday Self-Care & Wellness Intentions

Sunday Self-Care & Wellness Intentions

Ideas for Regularly Scheduled, Recurring Self-Care & Wellness Practices

Reflect on activities and strategies that you have enjoyed and that have seemed helpful. Are you starting to see any patterns? How about things that you've tried that didn't seem useful or that you did not like. Make notes to help inform your recurring self-care plan and activities.

Weeks 26 & 27—Time to Reconnect With Your Daily Practice

This is it! You're at the halfway point! You have now completed nine ethics topics. It doesn't matter if you arrive here exactly in Week 26 or if it took a few more weeks for you to get to this point. You are here, and we are proud of you! We are sure your to-do list is overflowing. We know you have so much on your plate and that each week we are probably adding more. So, take the next 2 weeks to just reconnect with your daily practice and work on those tasks.

Review & Reflect— How are you feeling about your efforts to build your daily practice? Sometimes, building daily practice can feel like a chore. We hope as you come to the halfway point, your daily practice feels less effortful and more fluid. Take a few minutes to reflect on the components we are asking you to implement (thinking about, noticing, talking about, acting on, and reflecting on). Are some easier than others? If so, why? Maybe try connecting your values to the components that feel wonky or more challenging.

Celebrate— That's right; time to celebrate your successes! Make that list and bask in the glow of how you are showing up to benefit your clients, supervisees, trainees, colleagues, profession, and yourself.

Refine, Plan, & Act— Take a minute to review your task list, and as you have done before, prioritize tasks you can get done in the next 2 weeks. You can use these weeks to clear a lot of smaller tasks. Alternatively, you might have a bigger project that you want to complete or get a hefty start on. There's no right approach; just take a thoughtful, planned approach to building out your tasks. And if there is a topic or a task that you have been putting off, maybe lean into that and try to identify why are you avoiding it. Can you find space on your calendar to tackle it soon?

Tasks for the Next Weeks:

____

____

____

Tasks to Find a Cozy Spot on Your Calendar:

____

____

____

Now's the time to get after your tasks. You probably have some lingering from past weeks, so use this week to get them done if you can. Move remaining tasks to your future task list and get them on your calendar. Slow and steady wins the race, so don't push too hard, just knock out one task at a time. You are doing really great! Think of the positive impact getting these tasks done will have on you, your clients, your supervisees, and the overall work culture.

Look Ahead— Maybe take time to scan not only the coming weeks, but also the rest of the biweekly pairs. There might be a specific situation brewing related to one of the topics scheduled for later that would be valuable sooner. If so, it's perfectly fine to jump ahead if needed. Just remember to come back to the weeks you skipped.

Inspiration to Reconnect— At this point you might feel that you have more questions than answers when it comes to ethics. And we feel that in our souls. And even though that feels uncomfortable, that means that you are really leaning into developing a deeper, analytic relationship with ethics topics and dilemmas. So, lean into those questions. Truthfully, being able to identify, ask, and explore the right questions is much more useful than thinking you have the answers. So, make friends with those questions, as they are the signposts that will help you navigate this journey!

Weeks 28 & 29—Assigning Tasks to Supervisees & Trainees

Week 28

Check in—Are you feeling well-rested and caught up? You are making such great progress moving through this book and developing your daily ethics practice. We are quite sure that you've had some challenges already, but we hope that some of the content is fun and that you feel like you are learning and broadening your perspective on ethics topics.

Focus for the Week—Let's talk about delegating tasks. As a supervisor of those accruing supervised fieldwork experience (trainees) or those implementing your behavior analysis programming (supervisees), you should be delegating tasks as part of your time management strategies and as a training tool. Seems simple, right? Not so fast. Delegating is defined as entrusting another, appointing someone as your representative, and assigning responsibility or authority (Merriam-Webster, n.d.-a).

When you delegate a task or responsibility to a trainee or supervisee, you remain responsible for the outcome! So, delegation requires that you are confident that the individual has the knowledge and skills necessary to successfully do the thing assigned. Not only that, but you also need to make sure that the individual is allowed to do that task, as some funding sources or company policies may not allow certain individuals (e.g., behavior technicians, trainees) to carry out certain tasks.

Values Check—As with Weeks 7 & 8 and Weeks 13 & 14, connecting with your values can be really helpful for topics that seem easy or like common sense. Which one or two of your values might be most useful in helping you connect to the topic of assigning tasks to your supervisees and trainees?

Risks & Benefits—Delegating tasks is a tricky thing, and the risks span from low to high. One reason task delegation is tricky is that we are likely to contact negative reinforcement once we hand over the task. This can lead us to delegate only our non-preferred tasks to others. Those individuals may come to resent us for passing along our "trash." Similarly, delegating tasks that are not well matched to the individual's skills can lead to feelings of frustration and distrust, as team members may feel like their skills are wasted or that they are unsupported and have to figure out things on their own. So, careless delegation can damage relationships and the work culture.

Delegating tasks outside of someone's scope of competence, or without proper support, can also invite risk of harm to the individual, client, and others. Consider delegating the task of conducting a skills assessment to a team member who has only assisted in running them. What could go wrong here? Yeah, you guessed it: inaccurate data and lost time at the minimum.

There is also a degree of risk in never delegating, as that can leave you overburdened and your team feeling like you are not invested in their growth or that you don't trust them. Finally, there is a risk of delegating unfairly (e.g., always assigning preferred tasks to some team members and non-preferred tasks to others), which can be demoralizing to your team.

Can you think of some benefits of thoughtfully delegating tasks? Here are some we identified: Team members feel valued, they increase their skills and knowledge, and you have time to get other things done.

Antecedent Strategies—Just reading this section so far is a good antecedent strategy, as you may already be thinking of tasks you should stop delegating or some tasks that would be appropriate for you to turn over to someone else.

It can be helpful to have a clear list of competencies for each position, a way to objectively evaluate those skills to measure competency (i.e., the ability to perform the task accurately and independently over time), and a way to track the competencies. Doing so will allow you to make data-based decisions about which tasks can and cannot be delegated.

When you are considering delegating a task to someone, especially the first time, make sure you have the time and resources to clearly explain the expectations and provide any needed support to ensure the task is carried out successfully.

To avoid unfair delegation of tasks, consider tracking who is assigned different tasks so that you can ensure that your team members equally share the burden of the less awesome tasks and equally share the benefit of the good stuff.

Spend some time exploring your implicit biases to help you proactively identify areas where you might be more likely to assign preferred or easier tasks in a way that is informed by your biases rather than the individual's skills and needs. If needed, head on back to Weeks 15 & 16 to reconnect with this content and with strategies to help you lean into this topic.

Red Flags—What are some indicators that you, or someone else, are inappropriately delegating tasks? Some red flags can show up in the work culture in the form of people expressing discomfort with tasks or lack of meaningful work, sharing that they feel unsupported, or avoiding working with certain supervisors. Poor outcomes, like tasks not being completed correctly or on time, or lack of client progress, can be indicators that tasks are being assigned inappropriately. Team member or client injury can also be an indicator of inappropriate delegation.

As supervisors, we might also notice that we always assign our non-preferred tasks to others. Finally, upon honest self-reflection, we might see that we tend to delegate preferred tasks or give more support of assigned tasks to team members who we perceive to be more like us or to possess certain qualities. In other words, we may find our implicit biases unintentionally, but still harmfully, creeping into our task delegation behavior.

Addressing—Alright, listen up, because this is important: You need to address how you delegate tasks, even if you think there are no issues here. Why? Because your delegation behavior is likely not purposeful, and you likely don't actively evaluate it. The easiest thing to do is commit to regularly taking stock of your behavior related to delegating tasks, and we've got some actions for you this week to help.

You also need to teach your supervisees and trainees about the importance of taking a thoughtful and purposeful approach to task delegation. More importantly, teach them what to do if someone delegates a task that they are not ready to take on, or if someone is regularly delegating tasks in a way that they perceive to be inequitable. Finally, if you're worried that a colleague or leader is engaging in the inappropriate delegation of tasks, gather some data and have an open discussion about what you are noticing, how they approach delegating tasks, and how their task delegation is impacting you (or how it could be impacting others).

Weekly Actions

MONDAY—THINK ABOUT IT

- Which of your core values are linked to appropriately delegating tasks to others? Can you think of ways that one or more of your values can help you attend to your delegation activities?
- Think back to when you were a trainee or were in a new position. Did anyone inappropriately delegate tasks to you (e.g., you could tell they were asking you to do it because they did not want to, they delegated without clear expectations, you had not yet mastered the skill or were not confident, you felt their delegation was not equitable)? How did that make you feel? What was the outcome?
- Think about how you delegate tasks. Do you always have objective, measurable data to indicate that the person has the skills necessary to successfully carry out the task? Can you identify any red flags in your behavior or in that of your supervisees or trainees?
- Make yourself comfy, and think about your implicit biases. Remember that we all have them—as we discussed in Chapter 3 and Weeks 15 & 16—and that's okay. It just means you are human, but you're doing the important work of making space and time to learn about yourself and taking action to grow.

  If you are open and honest with yourself, can you identify times when you engaged in preferential delegation? How do you think that made the other person feel? Does that behavior align with your values?
- Jot down notes and thoughts:

TUESDAY—LOOK FOR IT

- Does your workplace have clear job duty descriptions and measurable competencies? Are there antecedent strategies in place for evaluating competencies? Are there requirements for demonstrating competency before being delegated a task?
- Can you identify any instances of tasks being delegated inappropriately to you or others?
- Does your workplace provide training to behavior technicians in self-advocacy strategies so that they can speak up when they perceive that they're being delegated tasks without having demonstrated competency?
- Jot down examples:

WEDNESDAY—TALK ABOUT IT

- Get a conversation going about this content with a colleague, supervisor, or mentor. You might say: "How do you approach ensuring that you are delegating tasks appropriately?" Move the conversation forward by talking specifically about how to determine if an individual has demonstrated sufficient competency for a given task, and how to ensure that supervisors do not abuse their ability to delegate tasks and are equitable in their task delegation.
- Chat with leaders in your workplace about the policies or practices, in place or missing, related to this topic. Are things great? Take the opportunity to praise them! Maybe there is room for improvement on this topic and you can share some suggestions.
- Bring up this conversation with your supervisees or trainees. Ask them if they have ever been delegated something outside of their scope, and what happened. Ask if they feel that tasks are delegated equitably. Ask them how they would respond if you or another supervisor asked them to do a task without first ensuring that they could demonstrate the necessary skills. Ask them to describe one thing you could do right away to improve your delegation practices.

- Jot down notes and thoughts from your conversations:

THURSDAY—ACT ON IT

- Make a list of the tasks that you regularly delegate and to whom. Are there any patterns (e.g., you delegate your non-preferred tasks or tasks you are not comfortable with, you tend to delegate preferred tasks to your favorite team member)? Evaluate the outcomes of tasks you delegate (e.g., are there often issues with accurate, independent, or timely completion)?
- Take a few minutes to write out your process for determining competency before delegating tasks. Do you feel confident that this approach is successful and ethical? Are there areas for improvement? For example, perhaps you need to identify high-risk tasks (e.g., assessing and intervention testing for dangerous behavior, communicating with caregivers regarding things like consent) that require a more structured evaluation of team members' competencies before you delegate the tasks. Commit to making a plan to implement the improvements you noted (e.g., put it on your calendar or to-do list, share your goal with a colleague or supervisor).
- If you identified any red flags with yourself, colleagues, supervisees, or trainees, make a plan (with due dates) for how to address them using some of the strategies provided or that you came up with.
- Jot down thoughts and tasks:

FRIDAY—REFLECT ON IT

- Compare how you thought about delegating tasks at the start of the week versus now. Do you feel that you have a better understanding of why it is important to have a structured approach not only to delegating tasks but also to evaluating your delegation behavior?

- Spend some quiet time reflecting on how implicit biases may have impacted you when you were a trainee and how they might show up for you as a supervisor. As you identify your implicit biases, how will you make space to notice when they might be unintentionally impacting your delegation behavior?
- How will you make sure that you carry out your planned activities to improve how you delegate tasks?
- Jot down something new you learned or something that surprised you:

Resources to Check Out

- In the *Harvard Business Review* article "To Be a Great Leader, You Have to Learn How to Delegate Well," Jesse Sostrin (2017) describes four strategies for leaders to evaluate and lean into purposeful delegation.
- For some great tips on pre-planning, *wh*-questions to help you make decisions around delegating, and strategies for how to delegate, check out the article by Jin and Merritt (1993).

Week 29

Check in—Hey there, rockstar! So, after spending a week focused on ethics related to delegation, you should have a pretty good feel for whether you're an over- or under-delegator. Or maybe you're like Goldilocks and your delegation is generally "just right." But maybe you are susceptible to hastily turning things over, or white-knuckling tasks in certain situations, like when you're really stressed out. Either way, it's okay because you've got this week to get some tasks completed and list out future tasks to ensure that you continue to engage with this ethics topic regularly. As always, there's a scenario for you to check out too!

Scenario—Read the scenario related to delegating tasks, and see what you think. You spent last week reflecting on your history and current habits around task delegation. So, take that analytic approach here, and see if you can notice any risks, red flags, proactive strategies, or actions taken. Ask yourself what worked, what didn't, and what could be improved for next time. Remember, our focus is not on making sweeping determinations of "ethical" or "unethical." Instead, we're working to practice the subtle and complicated act of noticing, predicting, and analyzing those moments where we can improve our behavior or the environment to minimize harm and maximize good!

> *Janice is a trainee who is just beginning to accrue her supervised fieldwork experience hours. She likes her supervisor, Dalton, even though he is frequently really busy and doesn't quite have the time to fully explain things or give her the chance to practice after he demonstrates something. She knows that she is smart, and she takes it as a compliment when he asks her to complete a task for him, even when she has to figure things out herself because he is stretched so thin, and she is happy to help.*
>
> *Plus, one time she mentioned that she was a bit uncomfortable when he asked her to run a preference assessment after he just described the procedures, and he told her that part of the job is being able to figure out how to implement things from written or vocal descriptions and that he wouldn't ask her to do something unless he felt she could do it. Recently he asked her to write up several goals and programs, even though she has never helped write goals or programs, let alone done it on her own. She figured that there wasn't any harm because it was just for skill acquisition and not for behavior targeted for reduction.*

- **VALUES CHECK**—How does this scenario sit with you? Can you identify things that Dalton did (or didn't do) that don't quite sit right with you? How does his behavior align or not align with your values?
- **IDENTIFY RISKS**—Circle language indicating a risk that was present, and then write down the risks/resulting harms.

  Did Dalton or Janice also identify the same risks? Did they take action to minimize or avoid the risk? Are there other risks or actions you can think of that could be related to this situation?

Be sure to consider specific aspects of the scope of risks or harms!

- **IMAGINE ANTECEDENT STRATEGIES**—Antecedent strategies are not going to show up in every scenario. Draw an arrow to indicate any places where an antecedent strategy could have helped. Maybe you can think of some from your past or current work experiences. Now make a list of those antecedent strategies, and briefly describe what it would look like to implement them and how you would know if they were effective.
- **IDENTIFY RED FLAGS**—Underline any red flags that you see, even if Dalton or Janice didn't notice them. If the person didn't notice them, write down what factors may have contributed to their ethical blindness (e.g., time constraints, lack of knowledge, authority, common practice, perceiving that there was no risk, lack of oversight). Can you think of any others? Go back through and draw an arrow to indicate any places where an antecedent strategy could have helped, and then briefly describe what Dalton could have done as the supervisor.
- **IMAGINE INSTANCES OF SOMEONE ADDRESSING THE ISSUE**—Not all scenarios will include a description of someone addressing the concern or issue. Put an * next to a place where you think Dalton or Janice could have taken action to address the issue, and describe the action that should be taken and how they would know if it was successful or not. What would it have looked like if Dalton realized the issue and took action to fix it? How about if you were Dalton's colleague—what would you say to him? What if Janice was the one to address the situation? What could she have said?
- **CONNECT**—Have you ever been in a similar situation or known someone who was? What was the outcome? What would you do differently, or recommend the other person do differently, now?

Taking Care of Business—What are your tasks going to be? Maybe you just need to spend time getting the tasks from Thursday done, and that is A-OK! Or maybe you have a few new tasks, like adding a row to the spreadsheet you use to track your trainees' progress or techs' skills that will allow you to track the date you ensured competency before delegating a task. Whatever your tasks are, be proud of yourself that you're taking this week to connect with this ethics topic and make some improvements.

Words of Inspiration—Being a supervisor and leader is such a wonderful opportunity to give others the chance to learn and grow when you hand over well-aligned tasks to them. That also frees up your time to work on other tasks that only you can do. So, no matter where you find yourself at the end of these 2 weeks, having to take some tasks or give others the opportunity to take on a task, you are doing great!

Weeks 30 & 31—Addressing Professional Boundaries & Multiple Relationships

Week 30

Check in—You are cruising through these weeks and are likely in a good groove—keep up the good work! We are excited to engage in this next topic with you, which is a tough one, so get comfy and here we go!

Focus for the Week—Our topic over the next 2 weeks is a big one and may cause feelings of unease and discomfort, as we have all dabbled in potential or actual multiple relationships. None of us are exempt from this one because we are humans who work with humans, and humans are complicated! We'll focus less on the right and wrong, or good and bad, and more on how we can successfully identify and navigate these complexities. We want you to lean in with us to think, talk, and act on the hard things.

So, what are the hard things that we are talking about? Well, as behavior analysts, we have a responsibility, first and foremost, to our clients. But we do not work or live in a bubble. We work closely with families and other professionals to provide compassionate care and supervision. The very nature of our work can make it challenging to establish and maintain professional boundaries and can sometimes even increase the likelihood of entering multiple relationships. When we enter into a multiple relationship, we move further away from client- or supervisee- centered care. So, we want to know how to avoid multiple relationships, how to identify them, and what to do when we find ourselves (or someone we supervise) in one.

Let's build a foundation by clarifying some terms: conflict of interest, multiple relationship, and exploitative relationship. A conflict of interest occurs when a behavior analyst's professional interests intersect with their personal or other professional interests, potentially compromising their ability to make decisions (APBA, n.d.). For example, say the owner of an ABA company has a child who wants to pursue certification as a behavior analyst. The individual reaches out to their team of behavior analysts to see if any of them have the capacity to take on a trainee. This is a potential conflict of interest, as it may cloud judgment and could impact service provision and supervision.

So, what exactly is a multiple relationship then? It is when we, as a professional, take on one or more additional roles with someone we are responsible for professionally (e.g., client, caregiver, supervisee, trainee). These additional roles can be thought of as boundary crossings, where we move from our professional role into some other role (personal or professional) that may not center on the safety and needs of the person we are responsible for. Additional roles or connections may arise before the professional relationship is established, during the relationship, or after the relationship has ended, and could potentially even occur without our knowledge or awareness. For example, if the organization you work for hires a parent of a client whom you are directly serving or hires a technician who is also the babysitter for your own children.

Lastly, we have exploitative relationships, where one person who is in a position of power uses their position for personal or professional benefit. Focusing on professional boundaries and multiple relationships is plenty challenging without complicating things even more for us right now. So, we'll dive into more advanced content on multiple relationships in Weeks 38 & 39 (power imbalances) and Weeks 46 & 47 (romantic or sexual multiple relationships).

Here is a high-level graphic to give you a bird's-eye view of what we are talking about. Our goal here is to help you develop the awareness and repertoire to successfully identify, assess, and address your course of action related to multiple relationships.

1. *Maintain boundaries to minimize the likelihood of a multiple relationship developing.*
2. *Assess the potential or presence of a multiple relationship.*
3. *If identified, take immediate action: Develop an action plan that includes steps for immediate resolution of the relationship.*
    a. *If resolution is not possible, the action plan documents safeguards and monitoring to manage the multiple relationship until it can be resolved.*

Now that we know what a multiple relationship is, it is important to understand the different ways in which these relationships can develop to increase your awareness and ability to identify them and respond accordingly. There are many types of multiple relationships, and the APBA document that we mention in the "Resources to Check Out" section will provide you with a solid rundown (APBA, n.d.).

Many of these can be consolidated into two pairs of avoidable versus unavoidable and expected versus unexpected relationships. The first group refers to the ability to control our engagement in a multiple relationship and either engage in steps to proactively avoid, or to manage those that are unavoidable. An example of an unavoidable multiple relationship may be if an individual you are providing services to is also a coach for your child's soccer team, or perhaps a bike instructor at your gym. In terms of expected and unexpected, expected is when you can see the multiple relationship coming. An unexpected relationship might be if your organization hires a technician who is also a parent to a child you are serving, or perhaps previously babysat your child.

Which types of multiple relationships have you experienced? Which are you most susceptible to at this point in your career?

Values Check—As we've called out before, you're likely in this profession and reading this resource because you are a dedicated, compassionate behavior analyst who is committed to the success of others. What values would you identify with working to avoid and resolve multiple relationships? As discussed in Weeks 5 & 6, we heavily invest in the development of trusting relationships with caregivers to ensure effective therapeutic rapport and positive outcomes. Take a moment to consider that your values may conflict with your role and responsibility as a behavior analyst, which places you and others at greater risk, although your intention is likely the exact opposite.

Let's think through an example: you value compassion and kindness and are serving a client who does not have reliable transportation. Your client has made incredible progress, but his mother just lost her job; is a single parent; and will need to discontinue services because she no longer has a car, and her home is not suitable for home-based care. So, you offer to transport her child to and from services, which is outside of the services that you are contracted to provide and may even result in liability to you and your organization.

You see, multiple relationships are not always about someone doing something bad. Things that feel "right" and "good" can simultaneously be problematic when it comes to our professional ethics. We'll spend more time discussing how to evaluate and navigate these scenarios, but take time to self-reflect and work to objectively define your values and potential conflicts in the context of this topic.

Risks & Benefits—The risks here are endless. Let's be real. Humans are complicated. We each enter relationships and interactions with our own histories, biases, preferences, expectations—you name it! Professional boundaries are in place for very good reason. What is the function of your behavior if you find yourself in a multiple relationship? It may have had nothing to do with you and just be happenstance, but how you choose to manage it is on you. So, is the function of your behavior to fulfill your responsibility as a supervisor or care provider, or is it to fulfill your own needs (e.g., contact reinforcement, build a side business) (Auteri, 2014)?

The risks of multiple relationships that are ineffectively managed, or not acknowledged and managed at all, can be damaging, extensive, and have a ripple effect. You could cause harm to others who you intended to help. You may increase their vulnerability, especially since you have a professional responsibility to them first. You could damage the relationship and future access to care or supervision. You could cause irreparable damage to their well-being and psychological safety. The behavior you model as a leader also teaches others what is acceptable for them to imitate. That is the ripple effect.

The other side of this is just as powerful, if not more so. The benefits of minimizing and effectively navigating multiple relationships create a safe space in which you and those you are responsible for can give and receive to access the intended benefits of the relationship. You can focus on the reason we are in the professional relationship in the first place, which is fundamentally our commitment to place the client, supervisee, or trainee first. Another reason is our desire to share our knowledge and expertise in service of others. Everything else is secondary.

Antecedent Strategies—Some of the strongest antecedent strategies happen at the organizational level. Having clear policies that focus on minimizing the development of multiple relationships is a great place to start. These policies might address things like prohibitions on providing any additional services to clients or caregivers outside of ABA services (e.g., no babysitting, no transporting clients, no taking on current clients or caregivers as clients in other businesses like personal training or massage therapy). Policies might also address maintaining professional boundaries and setting clear expectations around non-work-related activities with clients, caregivers, supervisees, and trainees.

Organizations should have robust training on the policies that should include scripts for how to address situations that are likely to arise with team members in different roles, as the way a *technician* should respond to a caregiver's invitation to attend a weekend trip is different than the way a *clinician* should respond. Scripts can be so helpful because many of us don't know what to say when faced with some of these situations that present possible multiple relationships.

Finally, having resources that can be used to assess and plan to manage (especially when the multiple relationship is unavoidable) and resolve a multiple relationship proactively minimizes multiple relationships going unaddressed.

We've talked a few times about how self reflection is a superpower, but it really is. Its sidekick is awareness! Relationships and people are complicated, and it can be very easy to find yourself in a multiple relationship before you even realize it. To avoid that, we use self-reflection and awareness to implement the most important strategy of all—the art of maintaining professional boundaries.

Individuals you are responsible for (e.g., caregivers, trainees, supervisees) may be very likable and may even share common interests (e.g., playing a sport, going to concerts, online gaming) with you, and it can be tempting to engage in those common interests. This could be your "quicksand," which we refer to as the variables or context that could make it more difficult for you to live your values and uphold your ethical responsibility. Quicksand can pull you in and get you stuck to the point where it is pretty tough to get out.

A great strategy to avoid this is to have a set of rules. These should be simple and nonnegotiable rules that you decide for yourself outside of situations in which there may be risk, so you can stick to them and minimize any vulnerability to those involved.

Think about what your quicksand is, as you consider which rules make sense to you. Maybe it is social situations and work settings outside of the work environment, like a conference or drinks before a business dinner. Set professional boundaries and rules. Maybe you decide that you won't drink alcoholic beverages in the presence of trainees or supervisees, or you'll set a maximum number of drinks. Maybe you set the rule that you won't stay at social events with supervisees and trainees longer than 1 hour.

Maybe you are working to build a side business and are tempted to share that with your supervisees or caregivers. If so, you might create a rule that you don't talk about your side business with, or in earshot of, anyone in the work setting at all, or at least not with people you are responsible for. Maybe you, or a supervisee, really value being of service to others and run the risk of agreeing to support caregivers or others in a way that might cross boundaries and invite a multiple relationship (e.g., babysitting, loaning money).

Whatever your quicksand guidelines are, they should be individualized and tailored to areas that you know are risky for you based on your self-reflection and degree of awareness. And yeah, honestly identifying and describing your quicksand doesn't feel great. But it feels way better than harming someone because you allowed a multiple relationship to develop or go unchecked.

Finding that you have some quicksand doesn't necessarily mean that you are in a multiple relationship, but it is critical to ensure that you engage in consistent self-reflection and have a keen awareness to know what you are even looking for. We think of this as scanning or monitoring for the potential development of situations where we serve in two different roles that may conflict and present risk. Again, the problem arises when there is a shift that threatens the client- or supervisee-centered model. And you might not be sensitive to this shift if you are not constantly watching for it.

Another way to think about this strategy is that you have a playbook for your primary role; maybe you are a supervisor and have a playbook for how you behave and engage with your supervisee. Well, if you think you need another playbook or set of rules for how to engage, when to engage, and what to discuss with that supervisee in certain situations (e.g., like when alone or not at work), you might have crossed into a multiple relationship. So, be aware; be reflective; and lastly, and most important, know your boundaries and maintain them.

Red Flags— We've dabbled in identifying some red flags already, like when we mentioned failing to effectively implement antecedent strategies and respond appropriately. Let's take a progressive approach to identifying red flags (Auteri, 2014). If you find yourself thinking *I'm an ethical behavior analyst and would never be in a multiple relationship*, then you've got a red flag that is creating a blind spot. Deep breath. Remember, it's not about good or bad here, but about developing a sensitivity to watch for those subtle shifts. Our first responsibility is to avoid multiple relationships, but the second is to identify them when they occur to effectively manage them.

So, the first red flag is if you do not engage in regular self-awareness and perspective-taking to evaluate and identify the presence or likely development of a multiple relationship. Indicators may include chatting longer than usual before or after a therapy session that is specifically unrelated to the services. Or perhaps your supervisee seems to be creating situations to increase the time that you spend together, especially alone. There may be increased risk when there are common interests that motivative you to engage in behavior and conversation that is outside of your primary professional responsibility.

There may be some areas of increased risk that are outside of your control, such as if you live and practice in a very rural area or area with very few practicing behavior analysts. The red flag, in those circumstances, is the absence of proactive discussion, awareness, and safeguards to effectively manage and monitor the multiple relationship. If you see that multiple relationships are present in your work setting (e.g., caregivers of active clients are also employees, clinicians are providing required supervision for their own manager or supervisor or for a family member, supervisors have close friendships with caregivers or supervisees outside of the work setting), well then, the culture that accepts these multiple relationships is the red flag.

The overarching red flag is not having a plan to watch for, identify, and address multiple relationships. Hope is not a plan, trust us. You need a plan! Without a plan, this can quickly snowball, and before you know it the secondary or multiple relationship now takes priority and impacts your decision-making and your ability to objectively fulfill your primary responsibilities. Those aren't just red flags; those are red flags on fire.

Addressing—Since hope is not a plan, let's get you a plan before you get burned. If you suspect or find yourself in a multiple relationship where you take on one or more additional roles with someone you are responsible for professionally, you can (1) assess and confirm, and (2) create an action plan that includes resolving the multiple relationship, if possible. This may include transferring the client to another supervisor or provider. If this involves a supervisory relationship, it may include transitioning supervision to another supervisor. This takes awareness, honesty, and self-control. We recommend that you work with a trusted colleague or mentor for support and guidance.

Ending the multiple relationship may not be possible, at least not right away. You may provide services in a small town in which there are no other supervisors available. Or perhaps there isn't another eligible behavior analyst available to provide supervision. This essentially means that you need a new playbook that captures the multiple relationship and incorporates safeguards to mitigate any risk until the relationship can be resolved. This requires an open, honest, and professional conversation, so pull out your difficult conversation toolkit.

You should define the multiple relationship, outline the risks, and describe safeguards that will protect everyone and help identify if any of the risks are presenting themselves. We recommend that you be specific and objective and include how you will know if the safeguards are no longer effective. Finally, you know we are going to talk about documentation! You've got to document all actions you take and what happened. Next week we'll talk about the importance of documentation, so you are getting prepped!

Similarly, if you see another behavior analyst or technician engaging in a multiple relationship, you'll likely need to have a challenging, but supportive, conversation with them and then facilitate, or help them develop, a plan to address the situation. Finally, if you see that multiple relationships are occurring in your organization, you can address it with a supervisor. You might open the conversation with something like, "I'd like to have a tricky conversation with you, and I am a bit nervous. I have noticed XYZ, and I'm a bit concerned that we might not be upholding our ethical obligations to avoid and actively manage multiple relationships. I might be missing something, so I wanted to have an open discussion with you."

Alright, lean in here and listen up. Maybe you are realizing that you've been in multiple relationships in the past that you did not address, or others did not address with you. Maybe you are seeing some of those red flags in yourself or your workplace. This stuff is tricky and uncomfortable. AND we know that you can do the hard things. We are here to walk you through this!

Weekly Actions

MONDAY—THINK ABOUT IT

- As you think and reflect, we encourage you to do so with no judgment. It isn't good or bad, it just is. And what matters is you have the right baseline to build from. You are doing so well by engaging and showing up.
- What was difficult to read or reflect on? Did any of the content cause a reaction of defensiveness? These may be areas to think further on and share with a trusted colleague to understand fully. Your self-reflection may be bringing some quicksand into focus.
- How do your values support you in committing to make space for the fact that you're at risk of finding yourself on the road to a multiple relationship with a client, caregiver, supervisee, or trainee?
- Did you find that your personal or professional values were in conflict with the professional ethical responsibility that we have in this area?

- What is your learning history in maintaining professional boundaries and minimizing the likelihood of multiple relationships? What about with addressing multiple relationships when you become aware of them?
- Did your supervisor discuss this topic with you? Did they model how to avoid, navigate, and address a multiple relationship?
- Have you effectively navigated a multiple relationship before?
- Can you make space to reflect on the fact that you can simultaneously be compassionate AND have clear professional boundaries? Think about how the ability to balance your professional boundaries may rest on *how* you address certain situations. Telling a caregiver that you are so appreciative that they value you and want to include you in a family vacation and that you cannot accept their generous offer because you value your professional therapeutic relationship and your ability to work with their child IS compassionate and kind.

  Telling a caregiver whose partner has suddenly passed away that you would like to accept their invitation to attend the funeral and that you hope they understand that you will attend only for a short while out of respect for the family members and friends who are there to support you IS compassionate and kind.
- Jot down notes and thoughts:

TUESDAY—LOOK FOR IT

- We briefly talked about quicksand, meaning traps that are hard to spot but are appealing to you and serve as reinforcers. These are areas that you are more likely to need a more structured script or plan for to hold you strong. So, look for your quicksand. Can you identify situations that you find reinforcing that might result in a multiple relationship (e.g., social situations with colleagues, interactions that provide attention and compliments, individual conversations with colleagues, spending time sharing information about behavior analysis with an interested family member)?

- Can you identify any risks in your present professional relationships and the boundaries that you have established or lack thereof?
- Do you have examples of boundaries that you or others have set? When boundaries have been set with you, did you appreciate them? For example, if your supervisor only stayed for the first hour or did not drink with the rest of the team, did that impact your perception of them?
- Are there policies in your workplace to prevent and address multiple relationships? Do you see multiple relationships occurring in the workplace? Do they go unaddressed, or is there a structured process for identifying and addressing them?
- Do team members receive training on identifying and addressing multiple relationships? Does your supervisor periodically check in with you about any possible emerging multiple relationships? Are there resources to support identifying, addressing, and resolving multiple relationships?
- Jot down notes and thoughts:

WEDNESDAY—TALK ABOUT IT

- It can be so healthy and valuable to share your thoughts and reflections with a trusted colleague, mentor, or supervisor. The process of talking through what we have worked on this week is a critical part of the process. It is also valuable for others because we are all accountable for protecting professional boundaries and minimizing the likelihood of multiple relationships. Be sure to share what most resonated with you, and why. What did you realize as you were reading through the content in terms of your degree of awareness, honesty with yourself, and how to set yourself and others up to be successful?
- Ask others about their experience in managing professional boundaries (if they are comfortable sharing, of course). What have they found to be successful, and what do they struggle the most with? What is their quicksand, and how do they stay out of it?

- Have a conversation with those in your workplace about existing or needed policies, training, or resources to support regular check-ins to identify and address multiple relationships.
- Begin an open conversation with your supervisees and trainees. Ask if they have had training on or experience with this topic. If your company has resources, do they know about them? Help them identify their quicksand. Develop scripts for how they can respond to situations that are likely to arise in your setting.
- Jot down notes and thoughts:

THURSDAY—ACT ON IT

- Take some time and reflect on your thoughts, observations, and conversations from earlier this week. Make a list of risk factors for yourself. These might be related to the quicksand you identified, or avoidance of addressing these types of relationships with yourself or others. You might also consider making a list of red flags and quicksand indicators for yourself.
- Think through and reflect on your current professional boundaries. Do you have guidelines or rules that you can readily identify that minimize the likelihood of finding yourself in a multiple relationship? Are there some you should develop? Take time to make an action plan (if you don't already have one) to activate if you see any of your red flags pop up.
- If you have reflected on this content and realized that you are in a multiple relationship, what steps can you immediately take to resolve the relationship? If this is not something you can immediately resolve on your own, make a plan to connect with a supervisor, mentor, or colleague and draft out contingencies and safeguards to protect yourself and the individual involved. What is your plan to address it with the other person, and what resources or supports do you need to be successful?
- What antecedent strategies, scripts, and rules will you develop, or work with others to develop, so that you and others in your workplace are not caught off guard in the future?

- Jot down notes and thoughts:

- Check out the *Multiple Relationship Assessment* mentioned in the "Resources to Check Out" section, and complete it to inform your action items and direction (APBA, n.d.)!
- Use the decision-making model by Afolabi (2015) that we mention in the "Resources to Check Out" section to evaluate multiple relationships across the three dimensions that are identified.

FRIDAY—REFLECT ON IT

- Was the content this week uncomfortable? If so, that's okay, and great job being honest with yourself. Can we share something? It's uncomfortable for us as we reflect on our past experiences and write about this topic. As you reflect and work through the discomfort, do you feel more confident and better positioned to move forward?
- Are there opportunities for you to share what you learned with others to expand the impact and share their perspective and impact as well?
- As you reflect on what you learned, challenge yourself one last time if you had thoughts like *I don't agree with that*, or *That doesn't apply to me*. Identify your blind spots and situations where you might default to an avoidance response. Share those reflections with your colleagues and supervisor to keep yourself accountable.
- How have you reflected on any discrepancies or conflicts between your values and behavior to increase awareness and mitigate risk? Revisiting our example from the beginning of the week regarding transportation, how might you respond to that differently now? Perhaps instead of transporting the client yourself, you may provide alternate resources that support the same outcome, without entering into a multiple relationship, and still maintain the safety of all parties.
- Jot down something new you learned or something that surprised you:

Resources to Check Out

- If you are (or become!) a member of APBA, check out the document titled *Multiple Relationships: Strategies for Understanding, Assessing, and Addressing* to specifically dig into the different types of multiple relationships (APBA, n.d.). We mentioned two primary categories above, avoidable versus unavoidable, and expected versus unexpected, but this resource outlines a more comprehensive list to ensure your awareness and preparation.

  Additional tools offered to members on APBA's website include checklists to evaluate the presence of multiple relationships, an action plan template, considerations for organizations and leaders, training checklists, and tips and resources for training.
- When you "Act on it" this week, use the *Multiple Relationship Assessment* from APBA (APBA, n.d.). This is a short and sweet check-in to evaluate where you are and guide where you go! In just seven quick questions, you can diagnose and take next steps.
- There is a great article from Auteri (2014) that outlines a thoughtful decision-making process to evaluate and consider the context of the multiple relationship, the purpose it serves, and other considerations to effectively navigate and manage through the gray areas.
- There's a great podcast episode of *Modern Therapist's Survival Guide* that outlines and explains the importance of awareness and having an ethical decision-making model (Widhalm & Vernoy, 2022). You'll hear examples and a discussion of considerations to help navigate a variety of different types of multiple relationships and potential responses.
- There is a great resource from Campbellsville University (2020) that expands on the gray area within multiple relationships and how and when to engage in them, while ensuring safeguards to mitigate risk. The article also describes the self-awareness strategies of "reflection on action" (reflecting after a client interaction) and reflection in action (reflecting during client interaction) to help engage in proactive decision-making.

- Take a look at the article by Afolabi (2015) for another great review of multiple relationships and a decision-making model. This model can be used in your assessment and review as you go over the three dimensions identified, including power, the duration or time of the relationship, and clarity of termination.

Week 31

Check in—This topic is big and can be complicated, usually because we make it so! We hope that you were able to take time to reflect and work through the action items to begin to develop greater awareness of professional boundaries and self-reflect on where you are today. You may have already developed a game plan, in which case, you are awesome! You may just be realizing that you need a game plan, in which case, you are also awesome! Either way, you're here and there's no better place. Thanks for your commitment to pursuing excellence and integrity in your practice with us.

Scenario—Read the scenario below and think through how to effectively navigate and manage the complexity that is us as people, and how to remain aligned to our ethical responsibility as behavior analysts.

> *Reese has been a behavior analyst for 5 years and recently completed her yoga instructor course. She provides services for a provider in a relatively small community that is in the same area as the studio where she is teaching yoga on the weekends. During yoga class one day, a parent walks in whom she knows well, as she provides ABA services to her son. Her name is Christina, and her son's name is Quentin. When Christina walks in, she is surprised and excited to see Reese. They have established a great therapeutic rapport, and Quentin is making excellent progress.*
>
> *Reese and Christina talk for a few minutes before class starts. Reese is feeling a bit unsettled, as she hadn't thought of this situation occurring. At the end of class, Christina approaches Reese and asks for her thoughts on the new technician who just started working with Quentin. She also asks if they can shift their start time to 5:30 p.m. instead of 5:00 p.m. on Fridays moving forward. Reese says*

sure and that sounds good, but feels a bit uncomfortable. Over the next several weeks, a few similar situations occur, and Reese becomes increasingly uncomfortable with the situation. She tries to change the classes that she teaches, but Christina then asks Reese what her yoga schedule was so that she could attend her classes, as she is a great instructor.

> ***Be sure to consider specific aspects of the scope of risks or harms!***
>
> - ***How big/bad is the risk/harm?***
> - ***Who is/would be impacted?***
> - ***Was the risk/harm possible, or did it actually occur?***
> - ***Is it likely that the actual or possible risk/harm will occur again in the future?***

- **VALUES CHECK**—After reading this scenario, which values did you relate with and think about? Can you identify any values that Reese may be struggling with?
- **IDENTIFY RISKS**—Circle language indicating a risk was present, and then write down the risks/resulting harms.

  What could Reese have done to minimize risk?
- **IMAGINE ANTECEDENT STRATEGIES**—What antecedent strategies should have been in place at different points in this scenario and throughout the relationship? Were there clear professional boundaries, and what were potential ethical blind spots for Reese?
- **IDENTIFY RED FLAGS**—Underline any red flags that you see, even if Reese didn't notice them. Was there self-reflection? Did she have awareness of the shifting dynamics in the relationship? Did she communicate and maintain professional boundaries?
- **IMAGINE INSTANCES OF SOMEONE ADDRESSING THE ISSUE**—Put an * next to a place where you think Reese could have taken better action and where she did take action with her supervisor. What was done well, in terms of identifying and discussing the issue and next steps? How could the issue have been avoided?
- **CONNECT**—Have you ever been in a similar situation or observed someone who was? What was the outcome? What would you do differently, or recommend the other person do differently, now?

Taking Care of Business—Take your time to work through the tasks, and if the only takeaway is greater awareness, that is a great first step. Sometimes the best progress is the small progress that is meaningful. Be intentional about your awareness and how it develops and informs your actions and behavior to maintain integrity and fulfill your ethical responsibility as a behavior analyst. Start small, and finish strong!

Words of Inspiration—Making space for this topic in your daily ethics practice is a bit like putting up speed limit signs or speed bumps. It reminds us of our continual ethical obligations and our professional guidelines, which are in place to ensure that we and others are safe and successful. You can push the speed limit, but you introduce risk to others who you are responsible for. Drive well.

Weeks 32 & 33—Locking Down Documentation

Week 32

Check in—You did a great job last week focusing in on how to identify, manage, and mitigate multiple relationships. Now, we are going to take a few step backs to take a broader view on a topic that spreads across just about every ethical requirement. Get ready to step out of the trees and see the whole forest.

Focus for the Week—Our focus this week may appear a bit drab and dry, but don't be fooled. This is one of the biggest pitfalls and oversights for many practicing behavior analysts—our ethical responsibility to create and maintain detailed and high-quality documentation (e.g., clinical, supervisory, financial, certification and licensure maintenance), regularly audit that documentation, and address any inaccuracies in a timely manner. And, if you supervise others, and you surely do, you are also responsible for training others to document their work efforts, to ensure the documentation is current and accurate, and to address and fix any issues.

As we talk through these responsibilities, the key here is accountability and ownership, as discussed in Weeks 7 & 8 in the context of supervisory responsibilities. These requirements are not "technicalities" and should not be afterthoughts. They are clearly our responsibility as behavior analysts, and we should treat all documentation activities with the same level of care and importance that we use when collecting client assessment and performance data. Did you know that our responsibility to document is found over 60 times throughout the *Ethics Code for Behavior Analysts* (BACB, 2020a)? It is pervasive because it is important!

We believe the following areas may present the biggest risk or are most often overlooked, but our list is certainly not exhaustive.

- **We have a responsibility to document fieldwork and ongoing supervision, including topics covered and mastered, feedback provided, and feedback solicited. There is also an expectation to evaluate our own supervision and document performance reviews.**
- **We have a clear responsibility to document services in our clinical work, including the conditions under which we will provide services, how to communicate a grievance, and a comprehensive and accurate depiction of our activities and efforts as part of each client's medical record.**

We have a responsibility to document our actions and the outcomes across just about every aspect of our ethics code. You name it—public statements, research, supervision, performance monitoring and feedback, discontinuation of services, referrals, professional activity, financial agreements, collaboration with colleagues, multiple relationships, scope of competence, and the list goes on.

The responsibility for documenting our work is ours, and ours alone. There are strategies and frameworks that can minimize risk and increase the likelihood of all the right documentation being collected (e.g., automated and electronic data collection, special positions in organizations to carry out certain documentation tasks), but ultimately, the ownership falls on us. Be sure to hold fast to that knowledge and awareness, and be mindful of how your documentation responsibilities influence your focus and behavior and impact others in ways you might not have imagined.

Values Check—When we take a behavior-analytic approach to this topic, let's be real, documentation tasks are not glamorous. There is response effort required and generally little to no direct reinforcement for completing many of these tasks. Or the reinforcement may be very delayed and all about negative reinforcement; about avoiding a possible issue. There are, however, tremendous harmful consequences for NOT complying with these ethical responsibilities.

How are your values aligned with documentation, billing, and accuracy of records? Do you relate to the values of ownership and accountability? Do you recognize your responsibility to uphold those values in this area, or have you made comments such as, "The intake team reviews that paperwork, those aren't clinical tasks." If this thought is familiar to you, we encourage you to review your values and the application in the context of your ethical responsibility to minimize vulnerability to yourself and others.

Risks & Benefits— Failure to deliver on your documentation obligations can result in harms to your clients, trainees, and supervisees. Failing to ensure accurate clinical data collection can result in you missing patterns that suggest there is a need for modifications to clinical programming, leading to stalled progress or the continuation or increase of interfering behavior. If clinical documentation is not effectively managed and maintained, other professionals who review the client's medical record will likely be limited in their ability to support the client due to poorly documented activities.

Do you complete your documentation in the way that you would want your care providers or supervisors to complete yours? Does the documentation that you complete accurately reflect your effort, focus, client progress, and plan? If not, this is a risk. If yes, this is a benefit!

Failing to maintain documentation related to caregiver communication and collaboration can make it difficult to evaluate the effectiveness of the therapeutic relationship. As we mentioned in Weeks 13 & 14 on the topic of developing and evaluating supervisory practices, dropping the ball with your trainees' or supervisees' documentation can result in delays in their ability to sit for certification exams or lapses in their certification and ability to provide services.

Another obvious risk of not taking the time to own these responsibilities is that you may face disciplinary action with certifying and licensure bodies, the organization you work for, and even the payor who is funding the authorized services. The contracts and agreements clearly outline documentation obligations for good reason; they provide objective, measurable evidence that you did the things you are supposed to do. We have a responsibility to ensure accurate billing of services, but if inaccuracies are intentional and unresolved, tremendous risk is presented to the organization, your continued ability to practice as a behavior analyst, and the client's ability to receive high-quality, effective treatment.

The worst-case scenario here is billing fraud, which can be described as intentional and inaccurate billing to increase the reimbursement received. As a behavior analyst, you might say, *Well, I don't bill for services, that's not part of my job. We have a billing department for that,* or *I use a third-party billing service, so they're responsible for any billing issues.* But not so fast. You are responsible for making sure that the data and documentation entered or passed along is accurate.

The benefits of completing, maintaining, and ensuring accurate documentation are honestly endless. There are direct and indirect benefits to the progress of and relationships with clients, caregivers, supervisees, and trainees. Upholding your documentation duties minimizes the likelihood of interruptions to clinical and supervisory services and your employer's ability to submit timely and accurate billing. Another benefit is that if you find yourself in a difficult situation where your professional conduct is called into question, well, the documentation you have maintained will play a critical role on the path to resolution.

Antecedent Strategies—Okay, now we are really getting to the good stuff! Yeah, we are nerds and love documentation! Don't judge. Maybe by the end of these 2 weeks, you will be a documentation nerd too! We have a responsibility to clients, families, supervisees, payors, our organization, etc., to document most all of our actions related to service delivery and supervision. Instead of sharing an exhaustive list of strategies for all scenarios, we want to provide you with a framework to consider and apply as you navigate your ethical responsibility to document, document, document!

- **Knowledge:** *Know the documentation requirements for the work that you are supporting, and know there is no excuse for not doing so.*
- **Environment:** *Ensure that templates and processes are structured in a way that simplifies our requirements and expectations. In other words, if you complete the template provided, you meet expectations! And you do it well.*

- **Training:** *Training! Templates are great to the extent that people know how to optimally use them. What is the frequency and quality of training to increase the likelihood that expectations will be met?*
- **Audits:** *Are the resources and training serving their intended purpose? In other words, is what you think is happening actually happening? Audits are necessary, because we are imperfect. They ensure that our efforts serve their intended purpose. That our templates and training and efforts minimize error and increase the accuracy and quality of our documentation.*
- **Review:** *How can the current structure be improved and ensure a high degree of confidence that any changes or updates are appropriately captured to ensure compliance with expectations? Create a committee or review process to ensure that templates and processes are optimally designed and functional.*
- **Feedback:** *Ensure there is a mechanism for users to share all of the things that are great and not so much! Solicit this valuable information often, to consolidate recommendations for improvement and action.*

Let's apply these strategies to our clinical services. How do we structure the environment to minimize risk to ourselves; our organization; and most importantly, our clients and supervisees? Given our responsibility to outline the services, scope, and process for resolving and escalating any grievance, be sure that YOU are the one responsible for reviewing prior to the onset of services. Be sure that YOU are the one reviewing the signed supervision agreement, how supervision will be structured, what to expect, and what is expected.

Here is another old adage: "If it isn't documented, it didn't happen." This should be documented and reviewed with families during the intake process and may be consolidated into one or multiple service documents. We encourage you to also review Weeks 40 & 41 related to consent and assent for services. If you are not sure the right requirements are in place, we encourage you to review the code with your supervisor in the context of the process and responsibilities that are outlined by your organization.

Is there a financial agreement, service agreement, client rights, and grievance resolution policy and process? Are you the one reviewing with the family and ensuring they have the opportunity to ask questions and have them answered? This is a space in time that is critical to the therapeutic relationship and should be used accordingly to establish expectations and smooth your path forward. Misunderstanding can be avoided by clear, proactive, and direct communication.

Let's shift to antecedent strategies that should be in place to support our responsibility to maintain detailed and high-quality client records.

Service Documentation

- *Do your service templates set you and others up to be successful?*
- *Are they structured in a way that captures the required components as outlined by the payor and organization?*
- *Or are there components that are lacking or unclear that introduce risk to compliance?*
- *Maybe the service documentation is far too long and excessive, which introduces the risk of error because of competing contingencies such as limited time or timeliness to convert services provided.*

Training

- *Is there clear training provided to document and verify services provided, or is there risk that you or another team member could complete the documentation incorrectly or render time that was inaccurate or not provided?*
- *Do team members have clear training and direction on how to escalate and change a services outline to match the services actually rendered?*
- *If you work for an agency, is there a clear zero tolerance policy for any intentional behavior aligned with billing fraud?*

Mistakes do happen, but we also have a responsibility to ensure those errors are captured and addressed from beginning to end. Is there a process to identify and escalate those errors to ensure proper documentation and resolution? How does a similar review look in the context of our responsibility to document supervision practices?

Red Flags—Have you ever taken over the responsibility of care for a client, supervisee, or trainee from another behavior analyst? The answer is probably yes. In your review of the documentation (e.g., treatment plan, fulfillment of services, clinical documentation, hours tracking, competencies evaluated), were you able to effectively understand what has been happening with the care (e.g., what has been mastered, what interventions were tried and the outcomes, what the next few steps are for goals and targets)? Was the documentation comprehensive and high quality so that you had a clear understanding, focus, and direction? If not, this is a red flag that lack of documentation could result in some ethical dilemmas with some pretty serious harms.

Is there a compliance team and audit process to review and evaluate client, supervisee, trainee, and other records? If not, you bet this is another red flag that issues could be arising and going undetected.

When it comes to documentation, identifying errors or issues early is critical to preventing the problem from growing. Is there a clearly outlined process to identify and correct any billing errors or supervision errors? Or do team members often say, "I rendered this time on accident, but I don't know what to do," or "I don't have my signed supervision contract." Uh oh; big red flags. Did you or others receive training on the importance of clinical and supervisory documentation? Many of the red flags in this area are identified due to the lack of antecedent strategies being implemented: incomplete or unclear documentation, lack of structure to ensure compliance, and lack of the appropriate training and resources that make it easier to do the right thing.

Addressing—We encourage you to critically review these areas, and we have outlined some considerations below. Ultimately, it is our responsibility as behavior analysts to address these areas, and in coordination with your organization if you work for one. We should always take the time to document, because we will almost always regret it if we don't.

Whether it results in ineffective care coordination, an incomplete or inaccurate medical record, a payor review, the loss of a payor contract, a disciplinary action for us or others, or some other impact on our ability to continue practicing within the field, we cannot afford to mess up documentation tasks. It is not a matter of if you will be audited or questioned, but when.

We want to engage in behaviors that support accurate and high-quality clinical documentation and have every degree of confidence that each client's medical record and supervisee's supervision reflects that level of diligence and care. So, if you are noticing that you might not be doing your best when it comes to documentation, that's okay! You've got us here to prompt you through some strategies for improvement. Maybe you are noticing some of those red flags in your workplace or in some of your coworkers. That's also okay!

Remember, the point of honing our ethics skills is not to make other people wrong or to point out errors. It is to best support safe and high-quality work. So maybe you need to have a difficult conversation with a leader in your organization. If so, take a deep breath. Then lean into preparing for a conversation that might start by you saying something like, "I have been thinking about our documentation processes. I think we do such a good job with XYZ, and I think there might be some room for improvement with ABC. Would you be willing to spend some time with me talking about this?" Either way, we hope that by the end of Week 33, you will feel like you have a good plan mapped out to address any needs you identify.

Weekly Actions

MONDAY—THINK ABOUT IT

- How do you think about your responsibility across these areas of documentation?
- Has your training included education and an awareness of your ethical responsibility to communicate and maintain documentation expectations across areas like clinical work, billing, supervision, and your own professional development?
- Are your values aligned with ownership of and accountability to these expectations, and does your behavior align with your val-

ues? Do you have the appropriate training and support to execute and ensure compliance?

- Jot down notes and thoughts. Frame your thoughts as to where you are and where you want to be and go with no judgment. Show up better every day, and start today.

TUESDAY—LOOK FOR IT

- Do you or does your organization meet the ethical responsibilities to review and document these areas before initiating clinical or supervisory services? If so, who is trained to complete this, and how well are they trained? Are they trained using BST to mastery? Do changes to documentation templates, processes, or expectations come with training and support to make the needed changes?
- Look at your templates (e.g., intake forms, service agreements, assessment reports, session notes, caregiver collaboration agendas, supervision agendas, supervision tracking). Are they designed to easily support detailed and high-quality documentation?
- What documentation is most time consuming, and is it necessary or could it be made easier to minimize the likelihood of errors and shortcuts?
- Are regularly scheduled audits and reviews being completed? What happens with outcomes of those audits and reviews? Have you or others received feedback on your clinical documentation that outlines what is excellent and what opportunities there are for improvement? Do you conduct your own audits?
- Do you meet with your supervisor to review required documentation in the context of service delivery, client progress, and supervisee and trainee success?
- Does your organization have a clear escalation team and process to identify and resolve any inaccuracies or conflict?
- Jot down notes and thoughts:

WEDNESDAY—TALK ABOUT IT

- Talk to a colleague, mentor, or supervisor about your reflections. What do you feel really solid with, and what makes you uneasy or less confident?
- Brainstorm together and solicit feedback. Based on your knowledge, what should be done differently and why?
- Speak with your supervisor about what you think is going very well and what opportunities exist.
- Meet with your expanded clinical team and/or supervisees to engage in perspective-taking based on the current structure of your process and expectations.
- Spend some time talking to your supervisees and trainees about the importance of documentation and the related risks and benefits. Ask them how they think the behavior analysts in the company feel about documentation. Ask them how they feel about it (e.g., what do they like/dislike, what is effortful/easy), and ask if they have ideas for improvements.
- Jot down notes and thoughts:

THURSDAY—ACT ON IT

- Review documentation that you regularly engage with (e.g., service agreements, consent forms, exchange of information forms, coordination of care forms, supervision contracts, fieldwork competency checklists, tracking systems for service and supervision hours), and ensure that YOU are fluent in using and training others to use them. Engage with the documentation systems more deeply by reviewing how you talk about and implement the documentation.

  For example, review the intake or service agreement for clients and use highlighting to note the key takeaways and talking points to review with families so that you are more likely to ensure they have the opportunity to hear, understand, and ask questions about the agreement. Review your service documentation templates and supervision contracts, and highlight or circle the things that seem most difficult for you to explain or others to understand. Then review the results with your supervisor and/or organization to see if they can be simplified or improved upon.

- Meeting with your supervisor or leader in your organization may sound something like, "I have been thinking about the great work that we do, the quality of our services, and the client progress that we celebrate, but the other half of that is ensuring there is accountability and ownership of documenting to account for that great work. I want to ensure that what we think is happening is actually happening and either gain greater confidence of it or make changes to be confident." You may also reflect and discuss how documentation can serve as a strategy to increase the likelihood for the right things to happen and minimize the likelihood of the unwanted things.

  Then expand to the other critical documentation responsibilities that we have across the code. This may sound something like, "Hey, let's take a look at the ethics code and find instances in which we have a responsibility to connect. Maybe we even use the search function and find the term 'document' throughout the code. Let's expand the group and share with our compliance committee to ensure that our responsibility is accounted for, and we feel solid across all aspects."
- Ask your compliance officer or committee to confirm the process for inaccurate billing and documentation or supervision deficiencies.
- Provide feedback or inform the training that is provided to technicians to ensure they recognize the importance of accurate and high-quality documentation and the risk of not doing so.
- Jot down notes and thoughts:

FRIDAY—REFLECT ON IT

- What content did we review this week that hit home the most? How has it changed your perspective, and how will it drive lasting behavior change?

- Are there opportunities for you to share what you learned with others to expand the impact and share their perspective and impact as well?
- As you reflect on what you learned, challenge yourself one last time if you had thoughts like, "I don't agree with that," or "That doesn't apply to me." Share those reflections with your colleagues and supervisor to keep yourself accountable.
- Jot down something new you learned or something that surprised you:

Resources to Check Out

- Check out the article by Luna and Rapp (2019), who created and used a checklist to improve documentation. These suggestions could also be embedded into an electronic template to serve the same function but minimize the need to reference a separate document as an individual completes their clinical documentation.
- Take some time to look through resources on the BACB website, including the *BCBA Handbook* (BACB, 2022b) and the video *2022 Supervised Fieldwork Summary* (BACB, 2021a). These resources are a great tool to understand our responsibility to document and to consider our ethical responsibility.

Week 33

Check in—Are you loving documentation now more than you did?? Yeah, we thought so. Your future self will thank you! This is one of those broader topics that applies to all the things, so let it marinate in your brain, and it will likely resonate more as you process and reflect on ideas and opportunities to make improvements and serve as a great model for others.

Scenario—We discussed the breadth of documentation expectations but will focus this scenario specifically on clinical documentation. We encourage you, however, to think through scenarios across all areas of the code and where your documentation is most misaligned. Again, this is not to discuss right and wrong or good and bad, but to minimize the risk of an ethical pitfall. Read the scenario below, and reflect on the importance of documentation and the impact of not meeting our ethical responsibilities to document.

> *Molly is a newly certified behavior analyst who recently started her first job as one. She's been through the company training and has assisted with intake and assessments at her prior company, but recently moved to Oregon and is about to complete her first assessment solo. Although she went through the initial training, she does not consider herself to be fluent with the documents, processes, and requirements. Because she's worried about looking foolish and unprofessional to her supervisor, she does not reach out for support and decides to figure it out on her own.*
>
> *She has an intake meeting with both parents of a new client, Matthew and Jason. Their daughter is 10-year-old Olivia. They have a wonderful meeting and spend several hours working through an understanding of Olivia's needs, her history, and the family's goals for Olivia. Molly asks the family if they reviewed and signed the intake paperwork, including the consent for services, client grievance, and financial agreement. They indicate they think they signed it electronically. They share that it was a little tricky to figure out in the system but say that they do not have any questions. Molly checks off the box on her intake outline for reviewing consent for services and moves on to discussing next steps and what Matthew and Jason can expect for the start of services.*
>
> *Throughout the next several months, Olivia makes rapid and wonderful progress while working with the technicians, and this is discussed during the caregiver collaboration meetings. During a home-based session one week, Olivia falls while outside on a break, severely injures her ankle, and begins crying. The behavior technician goes into the house to get Olivia's parents but cannot find them. The technician calls Molly, who then calls 911 to receive support.*
>
> *As the ambulance pulls up to the home, Jason pulls into the driveway and panics. He says he only left the home briefly to run to the store. The paramedics provide basic medical care to Olivia and rec-*

ommend that she get an x-ray, as there is significant swelling, and she may have broken her ankle. Jason takes Olivia to the doctor.

The next day, Jason calls Molly and shares that Olivia has a small fracture in her ankle from the fall. He questions what the technician was doing and how that could have happened. Molly feels quite overwhelmed and unsure of how to respond, as she thought the rapport with the family was excellent and the family should have known that they were required to be home during the session. After all, she thinks, it's clearly outlined in the agreement for services that a responsible adult must always be present. From Molly's perspective, Jason is the one who decided to leave the house during a session. Molly thinks, "How am I supposed to address and discuss this with the family? Jason is acting like this is my fault, but he signed the agreement."

Molly notifies her supervisor, Charlie. They review the intake documentation to ensure that it was signed, and they find it was not.

Molly finds herself horrified, and Charlie indicates that clinicians are expected to review the paperwork in detail, ensure that families have the opportunity to ask and have questions answered, and finally to also sign the agreement and check off on the checklist to ensure that the process is documented. Without that documentation, Molly and the organization are open to risk, which they will now need to navigate.

The family suspends all services, as their trust has been damaged and they are concerned for their daughter's safety. Olivia's parents, Matthew and Jason, ask for a complete copy of her medical record, including all documentation and notes from the sessions. They question the content of the clinical documentation, as it is minimal and not fully reflective of the progress that she has been making, which makes them question what else was not documented. They also question the training and experience of the technicians and Molly's oversight of her program. The family immediately discharges from services, stating that they are not comfortable continuing with therapy from the company and that they'll find another provider.

Charlie provided strong initial feedback to Molly, but quickly reflects on gaps the organization has in terms of training and ongoing audits that could result in newly hired BCBAs failing to understand and meet expectations. Molly has been a high performer since she was hired, and this incident highlights the importance of having the right checks and balances in place to minimize risk to clients, families, teams, and the organization. Charlie gathers a group to review the incident, to work to identify gaps, and to develop a plan to resolve any identified needs moving forward.

- **VALUES CHECK**—How did you feel reading through this scenario? Do you consider documentation to be an afterthought, or a key component to services and our ethical responsibility? Do you think that Molly values documentation, or perhaps she values client services but has a blind spot to the importance of documentation, given her history and training? What do you think are the values of the organization in terms of documentation and ensuring the team is set up to be successful? Do you have the awareness, training, and resources to effectively prioritize what you indicate is important to you?
- **IDENTIFY RISKS**—Circle language indicating a risk was present, and then write down the risks/resulting harms.

  Did Molly do anything to minimize the risk? Were there competing contingencies in place that could have been resolved to minimize the risk?

  Be sure to consider specific aspects of the scope of risks or harms!
- **IMAGINE ANTECEDENT STRATEGIES**—What antecedent strategies were missed that could have prevented the incident from happening? As we know, antecedent strategies increase the likelihood of a specific behavior occurring, which in this instance was to review the conditions under which services would be provided and the expectation to always have a responsible adult present. How could different antecedent strategies have been in place and documented, to ensure that Olivia's parents were home and participating during the session?

- **IDENTIFY RED FLAGS**—As we have discussed, documentation is a way to confirm understanding and agreement between parties to ensure that we function under the same expectations. Underline the red flags that you see. Molly missed the opportunity to document the conditions under which services would be provided. As a result, the family and organization were significantly out of compliance with expectations, which posed tremendous risk to all parties.
- **IMAGINE INSTANCES OF SOMEONE ADDRESSING THE ISSUE**—As you read through the scenario, use an * to mark where Molly should have taken better and more proactive action. She appeared to have great intentions and managed through a variety of new tasks, but what was missed? After the incident, how did her supervisor address the opportunity?
- **CONNECT**—As you have reflected on this content and the scenario, we hope that you are now a documentation geek as well. At a minimum, you recognize our individual role to ensure that we document and capture the amazing work that we do to support client needs. Talk with your supervisor, your mentor, and your colleagues, and take the opportunity to reflect on and ask if the infrastructure and processes that you support truly set everyone up to be successful.

Taking Care of Business—Take a look back at your notes and highlights. What was profound and impactful for you? What are you most proud of that you are already doing or have done? Be sure you share and celebrate it with someone! Perhaps you didn't put anything into place, and that is okay too. Take a moment to ask yourself why and what you can do to increase the likelihood of putting what you learned into practice. If you try to do it all, you won't go anywhere at all.

Words of Inspiration—Documentation is a gift to your future self and those who you serve. You will always be glad that you have it and regret it when you don't. It is not a matter of *if* it is needed, but *when*. Be kind to your future self!

Weeks 34 & 35—Resting & Reconnecting

Week 34—Time to Rest & Take Care of Yourself

Alright, three more biweekly pairs down! Hopefully you know what to do. Take care of yourself. Rest. Recharge.

Monday–Sunday Self-Care & Wellness Intentions

Plan out your intentions, and get after it.

Monday Self-Care & Wellness Intentions

Tuesday Self-Care & Wellness Intentions

Wednesday Self-Care & Wellness Intentions

Thursday Self-Care & Wellness Intentions

Friday Self-Care & Wellness Intentions

Saturday Self-Care & Wellness Intentions

Sunday Self-Care & Wellness Intentions

Ideas for Regularly Scheduled, Recurring Self-Care & Wellness Practices

Make a note of any practices you've tried or figured out that you'd like to continue.

Week 35—Time to Reconnect With Your Daily Practice

Welcome back from your rest week. Feeling good? If so, YEAH! If not, hang in there; you are doing great work! Just like in Weeks 10, 18, 26, and 27, take this week to catch up.

Review & Reflect— Do a quick touch base with yourself. This could be a good time to check back in with the work you did identifying or clarifying your values and biases. Do you need to refine your values? Can you identify examples over the past weeks when it was helpful for you to connect with your values? How has your understanding of your biases developed? Are you remembering to engage in self-compassion?

Celebrate— Yup, it's that time to shine your light on all the great things you've accomplished.

Refine, Plan, & Act— You are familiar with this by now, so get to planning and slaying! Oh, and remember to be honest about any of the biweekly pair topics or other tasks that you have consistently been putting off or moving to the back burner. How come? How can you lean in? Maybe try connecting the topic or task to one or more of your values.

Tasks for This Week:

Tasks to Find a Cozy Spot on Your Calendar:

Look Ahead— Flip through the next 6 weeks to get a feel for the next leg of your journey.

Inspiration to Reconnect— Take a minute to think back on the last ethics-related mistake you made. Don't cringe. Don't beat yourself up. Be grateful for that learning opportunity. You are doing the hard work of learning and growing. Thank past you for helping future you become stronger and more experienced.

Alrighty, you know the drill by now. This week is for getting back into the swing of things, clearing tasks, and getting yourself ready to start the next set of biweekly pairs. As always, be planful, but don't overdo it. And by all means, do have some fun!

Weeks 36 & 37—Transitioning Clients Responsibly

Week 36

Check in—Do you feel rested, reconnected, and ready to roll? We're going to start off with a topic that might surprise you a little. Have you spent time to think about our obligations with ensuring that clients' services are transitioned appropriately?

Focus for the Week—This week we'll tackle the ethical nuances related to transitioning a client from one clinician to another, regardless of if it is planned or unexpected, or temporary or permanent. Poorly executed transitions can result in interruptions to services that are problematic to the client's progress. There are a variety of reasons why a client may be transitioned to a new behavior analyst that could result in interruptions to services. What are some common reasons for client transitions that could lead to interruptions?

- A clinician must abruptly stop serving a client due to an unplanned event (e.g., injury, illness, or other unpredictable life event).
- A clinician has a planned event that requires them to temporarily hand over a case or cases (e.g., parental or familial leave, an extended vacation).
- A clinician leaves the organization.
- Existing clients are reassigned to a newly hired clinician to build the new clinician's caseload or reassigned due to changes in schedules or availability.
- A client needs a behavior analyst with the necessary skills in their scope of competence to effectively treat the client.
- A family has notified you that they will be leaving the state and will be transitioning to services at a location with the organization in their new state.

- A caregiver has requested a new clinician because they feel it is not a good match.
- NOTE: If a technician is not able to provide services, whether short term or long term, whether planned or unplanned, that can also produce interruptions to services. The clinician supervising the case should do their best to minimize those interruptions, so keep that in mind as we move through this content.

In any of these situations, a vital piece of the process is to ensure that there are minimal interruptions to services when a client transitions to a new behavior analyst and/or organization. It might be possible that you're a little surprised we are dedicating an entire biweekly pair to transitioning clients. And you also might be surprised that this topic has been separated from discharging clients or discontinuing services.

We (the authors) have found in our own experience that this is a topic that can use a little extra love, especially when considering ethical obligations. These types of transitions are very common in our clinical practice, so you should be spending some of your daily ethics practice thinking about and acting on this topic. And yeah, sometimes these transitions sneak up on you. We can't control if a clinician quits without notice or falls ill. But guess what? We are required to place the client's needs first, and we can control processes that allow for emergency coverage when life throws us a curveball.

Values Check—It's time to check out your values and select which ones relate to this topic. As we've already shared, it is possible that this topic might be a surprise to you, and maybe not one that you've spent a lot of time thinking about the ethical considerations with. So now is the time to identify which of your values best align with this topic. Which of your values seem most connected to ensuring that your clients' services are not stalled? Linking the topic to your values will help you stay connected to this topic, even after you've finished this biweekly pair.

Risks & Benefits—The risks associated with failing to effectively transition services for a client may seem inconsequential at first but can actually be very serious. If a client is transitioned without clear communication, a plan, and documentation, services could lapse, progress could be lost, and the client may experience other negative impacts (e.g., increases in dangerous behavior, exposure to unnecessary assessments or ineffective interventions). These interruptions to service delivery, even if only temporary, can also negatively impact the therapeutic relationship with caregivers, as their trust in the clinician and organization might be diminished, particularly if they are not active collaborators in the process. You might reconnect with the content and your notes from Weeks 5 & 6 and 21 & 22 on caregiver involvement.

A primary benefit of proactively transitioning services, whether short term or long term, is that you can actively minimize interruptions to services. Another benefit is that you'll increase the chances (to the degree that you are able) that your client will be able to smoothly transition into services with a new provider because you have provided the necessary information, documentation, and training (if possible).

If you happen to be on the receiving end of a case transition, taking a proactive approach allows you to gain the information necessary to begin to develop a strong therapeutic relationship and to be effective and efficient with the use of your time getting to know the client, caregivers, and programming. And if you are planful about the transition (e.g., have clear timelines and expectations, have clear communication), the process will also hopefully be less stressful for everyone involved, including you!

Antecedent Strategies—The first way you can set yourself up for success with transitions is to maintain your clinical documentation (go back to Weeks 32 & 33 for a refresher if you need to). This minimizes loss of time for your client in the event of an unexpected transition, as the clinician who covers or takes over your case will have a clear understanding of what's been done, what's happening now, and what the next steps are. We will also cover related topics in a few weeks (Weeks 44 & 45 on discharging clients and Weeks 48 & 49 on transitioning and terminating supervisory relationships).

The next strategy is to review the expectations for short/long-term and planned/unplanned transitions within your organization. If your organization does not have a policy and procedure for transitioning services, you might consider suggesting that it creates one. If you are a solo practitioner, you might consider creating resources for you to use in your own practice (e.g., policy and procedure, checklists, emergency protocols) that will support you in taking a structured approach to every transition. And should you end up working for an organization that does not have a policy and procedure in place, you can still use yours to guide you, and you can offer it to your leadership as a resource.

Remember, the obligation to make sure you transition a client appropriately is not on your organization or supervisor. They may be available to support you while you are completing the process and may even check in to make sure you are completing all the steps and see if you need help. But, at the end of the day, it is your job to make sure you are setting your client up for success with their new behavior analyst.

Also be aware of any relevant insurance, licensure body, or certification body requirements or expectations for transitioning a client. Have open conversations with your clients and their families about expectations at the start of services and about steps for a successful transition, should that be necessary. For example, let them know how much notice is best for them to give you, if possible, to help them transition to a new provider (e.g., they are moving, they decide to seek out services with a different provider).

If you're aware that you will be leaving your organization, communicate this with your supervisor as soon as you are able to start planning for the transition and identify who the next behavior analyst is that will be providing services. Or, if a client is transitioning to another behavior analyst and you're not leaving, still start planning for this as soon as possible, with the first step being identifying when the transition must be complete.

Red Flags—An overarching red flag is if you find yourself thinking, *I've never thought about this,* or *I have never discussed this topic before in any of my supervision, training, or practice.* One indicator that you might be at risk of slipping on facilitating a smooth client transition is related to how you think about transitions. For example, the following thoughts are red flags: "I don't need to prepare, everything is set," or "The organization is responsible for taking care of the transition or finding coverage if

I have an emergency," or "The next behavior analyst will need to conduct their own assessments and meet their organization's requirements, so I don't need to worry about it."

When might you be at risk for these statements to happen? Well, there may be times in your career when you are leaving a job because you disagree with aspects of the role, or maybe you find yourself burnt out or unhappy. Or maybe a family has surprised you and shared that they are switching to a new provider because they are unhappy with the services they are receiving. These contextual variables may impact your thought process when it comes to identifying an appropriate amount of notice to provide or how you communicate the steps needed to have a successful transition.

These are also red flags that you should openly discuss with your supervisees and trainees. Help them identify if they are in a situation where they may make a decision that will result in the improper transition of a client, and how to prevent that from happening.

Addressing— Maybe you're reading this and realizing that you have not thought about the importance of transitioning clients before. Maybe you are newly certified and have not yet had to transition a client. Or maybe you're identifying that you should have done more with a client you have transitioned in the past (either short term or long term). Take a deep breath. Remember, this is a judgment-free zone. If you are identifying that there have been or might be risks of violating our ethical obligations related to continuity in service delivery, the way to address it is head on. Take some time to figure out how and when you should talk about the importance of planning for transitions with your clients.

For example, in the first few weeks with a new client, you might take some time in a caregiver collaboration meeting to highlight the importance of getting ahead of short- or long-term transitions and what the process will look like in the event of planned or unplanned transitions. You might reach out to your supervisor and see what resources your organization has and what the current process is. You can also check out the resources at the end of this week.

If you identify that a client has been transitioned to you (either for the short term or long term) and it seems that the previous behavior analyst did not meet their ethical requirements for a smooth transition, then you'll probably want to look back at Weeks 19 & 20 on addressing the potential misconduct of others to help prepare for what might be a difficult conversation.

This also might be a conversation you need to have with your supervisor or leaders in the organization if you've identified a gap in the policies and resources provided by the organization. If you are finding that you or others are struggling to effectively manage short- and long-term transitions of services when planned and unplanned events occur, it is likely that your organization does not have a process in place to support behavior analysts with addressing these situations.

Weekly Actions

MONDAY—THINK ABOUT IT

- How do your values align with ensuring that you effectively transition your clients? How about with ensuring that you minimize the impact to clients when there are interruptions to service delivery? How can you use your values to help you stay motivated to ensure that you take all the necessary steps to ensure a smooth transition of services?
- Go back and look at the list of common scenarios in the focus of the week. Can you think of red flags related to each one? What about antecedent strategies or risks?
- What is your learning history related to transitioning services proactively for clients? What about taking a proactive approach to known upcoming interruptions to service delivery? Were you directly taught this, or was it modeled for you in your training and supervision?
- Spend some time thinking about if you can you identify times when you've been involved in the transition of services (e.g., the behavior analyst transitioning the client to another behavior analyst within the organization, the behavior analyst taking over for another behavior analyst). What about times when there were events that impacted service delivery (i.e., you or a technician were unable to provide services for a period of time)?

- Have you ever been on the receiving end of a poorly planned or executed transition? How did that impact you? The client? The caregiver?
- Can you identify risks or harms that occurred because you did not effectively transition services for a client? Can you identify what variables (e.g., not enough time, feeling burnt out, no policy and procedure in place) were occurring at the time that may have impacted your ability to appropriately transition the client? What about for instances when an interruption to services was not effectively managed?
- Do you know what needs to be included in a transition plan and what steps need to be taken to implement the plan?
- Jot down notes and thoughts:

TUESDAY—LOOK FOR IT

- Are there antecedent strategies in place for preparing you to proactively address interruptions and transitions for your clients in your work setting (e.g., organization policy and procedure, template for writing a transition plan)?
- Can you identify examples of where it could be beneficial to add resources and support to ensure you are proactively transitioning clients (e.g., onboarding, quarterly discussions with supervisor)?
- Does your organization have a documented process for transitioning clients? Do they have a transition plan template? If so, what is available (e.g., policy, procedures, checklist, scripts for discussing with caregivers, transition planning template, report template)?
- Jot down examples:

WEDNESDAY—TALK ABOUT IT

- Talk to a colleague, supervisor, or mentor about this content. Consider asking something like: "I have been thinking about making sure that services to clients are transitioned effectively to another behavior analyst and planning for interruptions in services. Do you think about that?" Ask them to share their process for transitioning clients or preparing for service interruptions. If they do not have an answer right away, ask if they would be willing to reflect on it and share it with you later in the week.

- Talk to your supervisor about your current processes, and ask for their ideas for improvements to your processes for transitioning clients, whether short term or long term.
- Talk to your supervisees and trainees about proactively transitioning clients and their familiarity with these topics. You can use the prompts from the week to help guide the conversation. You might have them make a list of the steps they already know to transition a client and review the list with them to identify where they may need additional training on the topic.
- Talk to your supervisees and trainees about the impact that service interruptions (both planned and unplanned) have on clients and steps they can take to minimize the impact.
- Jot down notes and thoughts from your conversations:

THURSDAY—ACT ON IT

- Make a list of the relevant documents, information, and training that would be necessary to transition a client on your caseload today. List out barriers that you might face when in the transition process (e.g., scheduling meetings with caregivers, scheduling time with the new behavior analyst). Got your lists started? Awesome! Now, do the same for times when you or a technician may be unable to provide services for a period.
- Take some time to formally review your process for transitioning clients. You might be reviewing your company's process as part of this exercise. You might review your supervisor's comments and ideas for improvement as you are completing this review. Does it address both short- and long-term transitions? What about for instances when the transition is unplanned (e.g., due to emergency)?

- Review the "Antecedent Strategies" section for this topic, and make a list of things you might need to review, revise, or create to support your effective transition of services and continuity of services. For example, this might be the time to create checklists or a transition plan template.
- If you identified any red flags in your review with yourself or your colleagues, supervisees, or trainees, make a plan and set due dates to address the red flags with the strategies provided.
- Review your processes for creating, maintaining, and storing your supervision documentation.
- Jot down thoughts and tasks:

FRIDAY—REFLECT ON IT

- Take a moment to reflect on the things you identified and the conversations you had. Do you feel like you have a better understanding of how to effectively transition services for your clients and why this is important? How about for ensuring continuity of services and minimizing interruptions to service delivery?
- Think back to the Monday reflection. How has your understanding of proactively transitioning services changed over the week? How has your understanding of planning for service delivery interruptions changed?
- What will you do to stay active with ensuring you have an effective process in place? What will you do to actively teach others how to proactively transition services and plan for service interruptions?
- Jot down something new you learned or something that surprised you:

Resources to Check Out

- The *Continuity of Services* toolkit (BACB, 2021b) and companion *Continuity of Services: Reminders for RBTs* (BACB, 2021c) document provide recommendations for navigating interruptions to service, including client transitions.

- Green et al. (2023) provide recommendations for addressing termination of service (including transitioning clients).

Week 37

Check in—Okay, you've got a week under your belt of specifically thinking and talking about proactively transitioning services for clients and how to prepare yourself for future transitions. You've also spent time thinking about how you can prepare for planned and unplanned events that may impact services. Maybe you've identified that there have been times when you have transitioned a client to a different behavior analyst or organization and did not complete all the necessary steps or have enough time to provide as smooth a transition as possible. Or maybe you've identified times that you did not set up a plan to have your clients' services covered while you were unavailable.

Reflecting on that may not feel great, but take a moment and acknowledge that now that you're aware of your ethical obligation, you can make sure to do better in the future. Now that you've spent some time adding this ethics content to your daily practice, take some time to review the scenario below.

Scenario—Take some time to read the scenario, and then go through the prompts. This situation might feel familiar to you, or it may be new to you. As you read through it, remember to push back on any tendency to make a knee-jerk determination about if there is a violation and what specific ethics standards are indicated. Take it slow. Engage all the critical thinking strategies you have been developing to engage on a deeper level with the information provided.

> *Marco has been unhappy at his job for the past 6 months. There has been a lot of staff turnover, and he feels like he is consistently being asked to cover for other people as they leave. He attempted to address this with his supervisor but felt like his concerns were unanswered. He has considered leaving the company but is worried that the issues making him unhappy will be the same at any company. However, he has hit the point of no return. Another behavior analyst just quit, and he has been asked to cover half of their caseload while maintaining his own without any additional support.*

With his concerns unanswered and an increased caseload, Marco cannot take it anymore and found a new job. He submitted his 2 weeks' notice and made sure all the programs, treatment plans, and all other information was up to date. He scheduled a meeting with as many families as possible but will not be able to meet with them all before he leaves. He did not do anything for the cases he took over from the previous behavior analyst because they should already be ready for transition since they were just transitioned to him. He has everything prepared to hand over to his supervisor, but his supervisor has not scheduled a meeting with him yet to discuss his transition. Marco is worried they are going to ask him to stay longer until they can hire someone new.

- **VALUES CHECK**—If you had to take a guess about Marco's values, what might one or two of them be, and how can you tell based on the scenario? Do they align with your values?
- **IDENTIFY RISKS**—Circle language indicating a risk was present, and then write down the risk/resulting harms.

  Did Marco identify potential risks? Did he take action to minimize or avoid the risk? Are there other risks or actions you can think of that could be related to this situation?

Be sure to consider specific aspects of the scope of risks or harms!

- **FIND ANTECEDENT STRATEGIES**—Go back through the scenario, and draw an arrow to indicate anything you think functioned as an antecedent strategy. Make a list of the antecedent strategies, and evaluate if they were effective at preventing things like risks or actual harms. How do you know? Could anything have been improved or added?
- **IDENTIFY RED FLAGS**—Underline any red flags that you find, even if Marco did not notice them. If he did not notice them, write down what factors may have contributed to his ethical blindness (e.g., time constraints, lack of knowledge, common practice, perceiving that there was no risk, lack of oversight). Can you think of any others?
- **FIND INSTANCES OF SOMEONE ADDRESSING THE ISSUE**—Put an * next to any language indicating that Marco took spe-

cific action to address the issue. Make a list of those actions, and then describe if they were effective or not, how you know, and anything you would have changed or added. Were there other points in the scenario where you think Marco could have taken a specific action to address the situation? If so what, what would that look like, and how would you know if it was successful? If Marco did end up having a meeting with his supervisor, how do you think it would go?

- **CONNECT**—Have you ever been in a similar situation or known someone who was? What was the outcome? What would you do differently, or recommend the other person do differently, now? If you were the supervisor, would you have done anything differently?

Taking Care of Business—It's possible that you now have a long list of resources that need to be developed. Or maybe you just identified some revisions and updates that need to be made. Maybe you ended up with a combo platter of some things that are stellar, some that need to be spruced up, and some that might need to be made from scratch. Whatever it looks like for you, use this week to work on some of those tasks to help set you up for future potential transitions.

You might review your caseload and determine if you have any clients that may be transitioning in the future or are in the process of transitioning services now. Hey, maybe it's the case that you have never talked about transitions with caregivers, and you want to make sure that you start those conversations as soon as possible.

You might have a long list, so if you're not able to complete it all this week, schedule some future time in your calendar to finish it. You might even loop in a colleague or supervisor to check in with as you complete the tasks to help hold you accountable, or better yet, to share the burden!

Words of Inspiration—We have codes, guidelines, and requirements, but at the end of the day, we have a responsibility to those we serve and the experience they receive. Ideally, we all hope to work for an organization that also takes this responsibility seriously, but ultimately, the responsibility is ours. We applaud you for taking the time to reflect and invest in this process. Your clients and their caregivers are lucky that you are growing in ways that place their needs first!

Weeks 38 & 39—Watching for Power Imbalances & Exploitation

Week 38

Check in—Hey there! Looking back on last week's content on transitioning your clients and other potential interruptions of service, how ya doing? Do you feel like you have a good plan moving forward, or at least know what steps you need to take to set up a solid plan for yourself (and hopefully some time on your calendar so you can keep working on it)? It's okay if you're identifying topics where you need to put in more work than for other topics.

Maybe you feel like you have a good handle on the ethical considerations about client transitions. Or maybe you identified that it is an area in which you need to invest time to strengthen your skills to ensure you're taking an ethics-forward approach with client transitions. Either scenario is okay!

Focus for the Week—We worked through addressing multiple relationships and professional boundaries in Weeks 30 & 31. Take a moment and grab the list of tasks you planned to work on (if you haven't finished them yet). You might find yourself adding to this list, or maybe you'll just make some updates to what you have already planned.

As we've already reviewed, a multiple relationship occurs when you have multiple roles with someone you interact with in your professional capacity (e.g., supervisee, trainee, client, caregiver, supervisor). When thinking about the risk involved with engaging in multiple relationships, it is important to acknowledge that such relationships could be exploitative and cause significant harm (Herlihy & Corey, 1992).

What makes a relationship exploitative or coercive? Well, this is likely to happen when there is a power differential in the relationship and the person with power uses it to gain something. First, let's talk about what we mean by coercion. Sidman (1993) defines coercion as "the control of behavior through: (a) punishment or the threat of punishment, or (b) negative reinforcement—the removal of punishment" (p. 75). In other words, coercion is threatening to or actually punishing someone to get them to do something you want them to do (Sidman, 1993).

And what do we mean by power differential? A power differential occurs any time there is a person with less power due to a hierarchical structure (Gibson et al., 2014). Think supervisee or trainee and their supervisor in a supervisory relationship; or in service delivery relationship, think client or caregiver and the behavior analyst providing services.

One specific example of a power differential you may have experienced is when two people start in parallel work positions (e.g., both are working as clinicians in an organization) and then one person is promoted to be the supervisor of the other (e.g., promoted to a clinical director position that supervises clinicians). This shift has moved them from equivalent positions to hierarchical positions and introduced a power differential or imbalance where one person now holds a position of power over their former colleague.

Another example is a supervisor having a romantic or sexual relationship with their supervisee, which is exploitative because of the inherent power differential between them. We'll walk through these situations (i.e., romantic and sexual relationships) in Weeks 46 & 47, so in this section, we're focusing on other potential power imbalances and exploitative relationships and their impacts.

Again, we provide you with a note of caution: If you've had experiences that might make this content difficult to engage with, please make sure you are keeping yourself safe. Reach out to a colleague or friend who can support you as you navigate this material and your self-care. If now is not the right time to work through this material, then skip ahead and come back when you're in the right space to consume it. Take care of yourself!

Values Check—It's time to check out that list of values you identified at the beginning again, as you start to work through this topic. The values that you identified for the previous biweekly pairs that discussed multiple relationships might be a good place to start. You've made it through a lot of topics at this point, and maybe you've added some new values to your list. Are any of them relevant to include here? Take some time to reflect on these values and think about how you can use them to guide you through this topic.

Risks & Benefits—What are the risks of power imbalances and exploitative relationships? Well, one biggie is that the person with less power may feel betrayed but also trapped in the relationship (Herlihy & Corey, 1992). They may feel as if they are unable to advocate for themselves, provide feedback, trust the person, or speak up when they disagree because they feel as if it could have negative consequences (e.g., impact the quality of services, make their supervisor upset). Another risk is that misuse of a power differential or an exploitative relationship could permanently damage the relationship, putting the services, supervision, or work relationship at risk.

A benefit of remaining aware of power differentials and refraining from engaging in exploitative behavior is that you'll avoid harming the individual and the relationship. Being aware can allow you to acknowledge that you are in a position of power and be mindful of how you build and manage your relationships. In doing so, you can also ensure that you are meeting your ethical obligations. Another benefit is that you will maintain the safety of everyone in your organization, including clients, caregivers, supervisees, trainees, and team members.

Antecedent Strategies—Alright, how can you get ahead of those risks and make sure that you are doing no harm and acting in the best interest of those you serve? The first step is acknowledging that you are in a position of power relative to those you are providing services (e.g., clients, caregivers) and supervision (e.g., supervisees trainees) to. Even if you think you are friendly and approachable, if you are in a relative position of power, a differential exists.

Second, take steps to ensure that you create active two-way feedback and regularly check to ensure that others are comfortable communicating with you openly about the services and/or supervision you are providing. This can be done with an open conversation, survey, or even check-ins carried out by your supervisor (i.e., they have meetings with your direct reports or caregivers).

Third, make sure others know who they can go to if they feel uncomfortable approaching you and are concerned about potential exploitative behavior. This might be your supervisor, someone in human resources, or a leader in the organization.

Fourth, continue to check in with yourself. How do you identify if you are using a power imbalance inappropriately? How are you ensuring that you are not coercive in your interactions?

And finally, it's a great idea to have a plan in place to address situations where you find that you may be inappropriately using a power imbalance and may be engaging in coercive or exploitative behavior.

You also likely have a supervisor with whom there is a power differential. Keep those communication lines open with your supervisor, and be sure to make them aware if you are finding that your behavior is impacted by the power imbalance (e.g., you're worried about giving them feedback, you are unwilling to provide alternative ideas or disagree with something they have proposed). And if you find yourself in a situation where you feel like attempts to address your concerns are ineffective, you might consider referring to the section on addressing misconduct in others from Weeks 19 & 20 and reaching out to a colleague, a mentor, or another supervisor or leader within your organization for support.

Red Flags—One example of a red flag is that you're finding that you're infrequently receiving constructive feedback from your supervisees and trainees. Or when you present an idea to them, they just agree with it and do not provide any considerations, feedback, or ideas of their own.

An example of a red flag you might see with caregivers or clients is they don't ask questions about the interventions in place, and they just agree with the information you share. And remember, while we are discussing this now, you might be thinking, *I'm aware of the power imbalance, and I've never had an issue.* But that thought should be a prompt for two things: (1) Think about how to keep this in your daily practice if you truly are in a place where you are aware of the power differentials at play and are actively making sure you are not engaging in coercive behavior, and (2) consider when the last time was that you evaluated your relationship with those you serve. If you've never evaluated this or have not in a while, now is a great time to ensure that you are, so you can make sure what you assume is happening (i.e., you've addressed the power imbalances at play) is the case.

Addressing—Let's spend a bit of time talking about what to do if you identify that you might be at risk of engaging in coercive or exploitative behavior or taking advantage of a power imbalance. The first thing that you can do is make note of the professional relationships you have that involve a power imbalance. Next, do a quick evaluation—have any of the red flags been present? If so, take direct and immediate action to address them.

You might consider having a conversation with the person to determine if you can resolve the situation yourself. You might reach out to a mentor or supervisor first to discuss it with them, or even ask them to be a part of the conversation to support you. If it has risen to the level that you will not be able to resolve it yourself, you'll need to bring the right people together, whether it's a supervisor or another leader in the organization, and work with them to determine if you can effectively continue providing services to the client or providing supervision to the supervisee or trainee. Remember, there is a power imbalance in *all* those types of relationships, but the risk comes into play when one person in the relationship starts engaging in behavior to take advantage of the other person, or when the imbalance is negatively impacting the relationship.

What if you become aware of another behavior analyst taking advantage of a power differential? Maybe it's your supervisor engaging in behavior that you feel is coercive with you, or someone has approached you with concerns about their supervisor or the behavior analyst providing them with services. Well first, pause for a second and gather some information. You can review Weeks 19 & 20 on addressing misconduct to help you prepare.

Remember, not saying anything in this situation can set you or others up for harm. And it is likely that the person may continue engaging in the behavior if you don't address it. If you are addressing it with your supervisor, you might prepare a description of the behavior, the impact it is having on you and your supervisory relationship, and maybe even some suggestions on how you'd like to see your supervisory relationship change. If you're addressing it with a colleague, you might take a slightly different approach and instead prepare some questions to get a little more information about the situation before deciding how to continue the conversation and raise the concerns.

Weekly Actions

MONDAY—THINK ABOUT IT

- How do your values support you in committing to acknowledge the power imbalances that impact your supervisory relationship and relationship with those you provide services to, and to address any misuse of a power imbalance?
- Take a moment to reflect on your supervision and training. What is your learning history related to identifying power imbalances, evaluating them, and addressing them if they have become exploitative or coercive? Did your supervisors discuss this topic with you?
- Think back to your previous professional relationships with power imbalances. Has there ever been an instance where you engaged in behavior that took advantage of the power imbalance, even if it seemed harmless (e.g., asking a supervisee to babysit for you)?
- How about in the supervisory relationships where you were the supervisee? Have you experienced a supervisory relationship where a supervisor took advantage of the power imbalance?
- Jot down notes and thoughts:

TUESDAY—LOOK FOR IT

- Can you identify examples of how power differentials are managed appropriately in your workplace? Can you think of any times when you've observed the misuse of a power differential in your workplace? What happened?
- Is there training available on how to acknowledge and manage power imbalances and how to address it if a coercive behavior is identified? Are there resources that could be helpful to make available?
- Are there policies to prevent misuse of power differentials? Is there a process outlined to address when instances are identified? Are the policies focused on protecting consumers of services and supervisees and trainees?

- Jot down examples:

WEDNESDAY—TALK ABOUT IT

- Have a conversation with your colleague, supervisor, or mentor about this topic. It's possible that having this conversation might be a little uncomfortable, but it could be great to start with something like "What steps do you take to make sure the power differential between us does not negatively impact our supervisory relationship?" Or maybe, if you've identified that there aren't any red flags in your supervisory relationship with them, you make note of anything you've noticed that they have done to acknowledge the power imbalance and to make sure it does not shift into coercive territory.
- Ask your colleague, supervisor, or mentor if they have addressed similar situations with other behavior analysts, experienced a coercive supervisor relationship, or, if they are comfortable, ask them to share if they have found themselves in a situation where they were taking advantage of a power imbalance.
- Review this topic with your supervisees and trainees. Start the conversation with acknowledging the power differential in your supervisory relationship and sharing the steps you take to address it. Discuss the risks involved with engaging in exploitative behavior and how to evaluate your professional relationships to watch for red flags and address them.
- Jot down notes and thoughts from your conversations:

THURSDAY—ACT ON IT

- Make a list of your professional relationships (e.g., clients, supervisees, trainees, supervisors), and identify any power imbalances involved. Evaluate these relationships and identify the steps you have taken to ensure open communication and to prevent any exploitation or misuse of the power differential. If you identify any red flags, make a plan to address them in a timely manner.

- Think back to your reflection and conversations from earlier this week. Make a list of any indicators you identified suggesting that you may be at risk related to these types of relationships (e.g., lack of training, lack of support and resources for addressing power imbalances that have resulted in exploitative behavior).
- Using the lists you created above, create a general action plan you can use if you identify any red flags. This might include a list of antecedent strategies to put in place and steps to take to address the situation.
- Review the related policies, trainings, and resources available in your organization for addressing this topic. Do they align with your ethical obligations? Are there any updates that need to be made? And if so, can you bring it up to your supervisor to make the needed updates?
- Jot down thoughts and tasks:

FRIDAY—REFLECT ON IT

- How are you feeling now that we are closing out the week? Were you uncomfortable? Were the conversations you had earlier in the week uncomfortable for either party?
- Do you feel more confident with how you might address these situations than you did at the beginning of the week? If you identified holes in your supervision and training related to this topic, do you feel like you have taken steps to start addressing them?
- What will you do to continue to evaluate these relationships, ensure that you continue to engage in alignment with your values, and address any concerns as quickly as possible to minimize negative impacts on the parties you have a position of power over?
- Jot down something new you learned or something that surprised you:

Resources to Check Out

- Wright (2019) shares a model from social work that includes questions to address power imbalances as part of a framework to assess cultural humility.
- In their article, Pritchett et al. (2022) discuss how power imbalances can be addressed with the goal of moving from a legacy of oppression to social justice. The authors provide guidance on how to begin to positively address power imbalances through perspective-taking, humility, and values-driven dialogue.
- Check out Week 30 for some other helpful resources on multiple relationships.

Week 39

Check in—What's your status? Are you feeling overwhelmed thinking about the power imbalances in your professional relationships? Take a deep breath. Do you know what's great? That you're recognizing the number of power differentials at play around you. That will allow you to be able to respond accordingly when interacting within or around relationships where power imbalances may start shifting toward the unethical end of the spectrum, where exploitation or coercion are present. This is the second biweekly pair touching on multiple relationships, and we hope this will prepare you for the upcoming topic on sexual and romantic multiple relationships and that you now feel empowered to detect and address problematic power imbalances in relationships, should they come up. Let's hop into that scenario.

Scenario—Read the scenario, and work through the prompts below. As a reminder, the goal is not to identify the specific ethics standard from a code that was violated—it might be possible that the scenario does not rise to the level of a violation. But keep an eye out for risks and red flags, as we reviewed last week.

> *Elijah and Claudia have worked together for many years. Their kids go to the same elementary school, and they often see each other at school events. Claudia was promoted last year to be Elijah's supervisor, and they feel like they have transitioned into their new*

relationship smoothly. They talked about how their relationship would change when Claudia first applied for the promotion. They have frequent check-ins to make sure they have open communication, including effective two-way feedback conversations. Claudia has been sure to tell Elijah that he can go to her supervisor, Riley, with any concerns about the supervisory relationship and they will support Claudia and Elijah with resolving any issues.

With her promotion, Claudia has been struggling to keep up with the additional tasks of her job and some of her personal tasks. One recurring issue is picking up her kids from school on time, a task she splits with her partner. Last month, when she needed to stay late to assist with a client emergency, she asked Elijah if he would be willing to pick up her kids. She figured it wouldn't be too big of a burden since they all go to the same school. He could just grab her kids when he picked up his own and drop them off at her house. Elijah was happy to assist.

Claudia is now having to take additional time to support a new team member and is tempted to ask Elijah if he would be willing to pick up her kids on a regular basis. She took some time and thought through the risks. While Elijah is her supervisee, they have a long working relationship and Claudia is confident he will voice his concerns with her if he feels like the request is exploitative. The risk of harm seems negligible, so Claudia decides to ask Elijah to help her out. Elijah does not hesitate and agrees to help her out twice a week.

- **VALUES CHECK**—What do you think Claudia's values as a supervisor are? Do they align with yours? What do you think Elijah's values are related to watching out for power imbalances? What about Elijah's values related to his relationship with his supervisor and coworkers?
- **IDENTIFY RISKS**—Circle language indicating a risk that was present, and then write down the risks/resulting harms.

  Did Claudia and Elijah also identify the same risks? Did they take action to mini-

Be sure to consider specific aspects of the scope of risks or harms!

mize or avoid the risk? Are there other risks or actions you can think of that could be related to this situation?

- **FIND ANTECEDENT STRATEGIES**—Go back through the scenario, and draw an arrow to indicate anything you think functioned as an antecedent strategy. Make a list of the antecedent strategies and who implemented them, and evaluate if they were effective at preventing things like risks or actual harms. How do you know? Could anything have been improved or added?
- **IDENTIFY RED FLAGS**—Underline any red flags that you find, even if Claudia or Elijah didn't notice them. If they didn't notice them, write down what factors may have contributed to their ethical blindness (e.g., time constraints, lack of knowledge, authority, common practice, perceiving that there was no risk, lack of oversight). Can you think of any others?
- **FIND INSTANCES OF SOMEONE ADDRESSING THE ISSUE**—Put an * next to any language indicating that Claudia, Elijah, or anyone else took specific action to address the issue. Make a list of those actions, and then describe if they were effective or not, how you know, and anything you would have changed or added. Were there other points in the scenario where you think one of the individuals could have taken a specific action to address the situation? If so, what, what would that look like, and how would you know if it was successful?
- **CONNECT**—What do you think would have happened if Claudia had reached out to her supervisor, Riley, for support? What do you think would have happened if Elijah had reached out to Riley? Have you ever been in a similar situation or known someone who was? What was the outcome? What would you do differently, or recommend the other person do differently, now?

Taking Care of Business—Did you modify or add any tasks from your existing list from the previous Weeks 30 & 31 on multiple relationships and professional boundaries? Use the rest of the week to continue to work on these action items, and be sure to schedule time in the future to finish anything you don't have time to finish this week. Remember, making these changes takes time. You're not going to be able to do it all at once. Taking a slow and steady approach allows you to continue to support your practices, and in the end, you'll be set up to manage these situations much more effectively and ethically. And don't forget to also schedule those recurring checkpoints to make sure you continue to engage with this topic in your daily practice and evaluate how you're doing.

Words of Inspiration—Power differentials in professional relationships will always be there. The fact that you are taking time to reflect on them, discuss this topic with others, take action to improve your own behavior, and model it for your supervisees and trainees is awesome and will have positive outcomes. You are continuing to protect those you serve as you take these steps, and you are also protecting yourself from potential harm. Remember that leadership is a privilege, and the power that you have should always be kept in check to align with the good and service of others.

Weeks 40 & 41—Ensuring Appropriate Consent & Assent Practices

Week 40

Check in—You've completed over two-thirds of the book! Congrats! By this point, you should be feeling like thinking about and attending to ethics topics and nuances is almost second nature. If not, it's okay; that shift will come. The last two biweekly pairs focused on client transitions and power differentials. That last topic is a doozie, particularly if you have a history of being taken advantage of by someone in a position of power over you in the workplace. You doing okay? If you find you need a rest, just remember that you control the pace. You can always take a week or two off for some additional self-care.

Focus for the Week—Let's spend the next 2 weeks thinking and talking about ethics related to consent and assent in your practice. Consent is generally defined as someone voluntarily agreeing to, or approving of, something proposed by another individual. Oh yeah, and the person consenting must have the capacity do so (i.e., be of a certain age and cognitive ability). Setting a threshold for who can legally give consent is meant to protect vulnerable individuals from being taken advantage of.

As behavior analysts, we are concerned with a particular version of consent called *informed consent*. Informed consent requires that the individual voluntarily agrees to something after they have been made aware of the risks of and alternatives to the proposed thing. The *Ethics Code for Behavior Analysts* (BACB, 2020a) provides us with specific factors to consider for obtaining consent related to services/research versus using or sharing information and specifies when we must first get consent from clients (or their legal guardians).

Assent, on the other hand, typically refers to those who cannot provide legal consent (i.e., children or those with cognitive abilities that might impair their ability to give consent). Assent is about the individual receiving services or participating in research indicating that they are okay with the activities. It may be more helpful to think about assent withdrawal as behavioral indicators that the individual is not comfortable with, does not prefer, or does not want to continue with the therapeutic or research activity.

Since this is not a textbook or a book specifically about the current BACB ethics code(s), we will leave it to you to make sure that you understand these concepts and obligations. Remember, we are focusing on increasing our skills to think about, notice, and proactively act with ethics content at the forefront of all we do as behavior analysts.

Values Check—We bet that all, or almost all, of your values apply to this topic. Where your values might show up in a helpful way over the next 2 weeks is if you find yourself rationalizing. For example, if you hear yourself saying, "But we've always done it this way," "There's just not enough time to make sure of XYZ," "We get a blanket consent at the start of services," or "My clients are too young or lacking in communication skills to give assent"; well, then visiting your values might shine a light on places you can lean into considering other perspectives and identify areas for growth.

Risks & Benefits—Certified and licensed behavior analysts and those conducting research clearly have professional risks for failing to obtain required consent, like disciplinary actions. Sure, that's not fun; however, in developing your ethics skills, it's much more important to start identifying risks for clients and others.

Failing to obtain informed consent can result in caregivers and clients participating in therapies and interventions that they don't want, that violate their personal morals and cultural practices, and that could cause physical and/or emotional harm. Imagine a caregiver not fully understanding the procedures they are asked to agree to that involve physical restraint for their child who has a history of physical abuse or who has medical conditions that could be worsened by being placed in a restraint (e.g., fragile joints, a shunt, asthma, or a heart condition). Such medical conditions might not be known to the clinician or might be overlooked as a significant risk factor. Failing to get proper informed consent can also erode the therapeutic relationship and trust, as the caregiver and client may feel powerless and even bullied.

The risks of failing to attend to assent and assent withdrawal from your clients include exposing them to aversive events and even pairing yourself and other therapeutic stimuli, including the general learning environment, with those aversive events. This can lead to increases in harmful behavior, as the client protests and attempts to escape the situation.

What are the benefits of always leading with obtaining informed consent and attending to assent and assent withdrawal? Protecting caregivers and clients, honoring their autonomy and agency, strengthening the therapeutic relations, and ensuring that the interventions and therapies are likely to be implemented in the natural environment.

Antecedent Strategies—A good starting place is to make sure that there is a policy clearly listing when informed consent is needed, who is authorized to carry out consent procedures, how it should be documented, and when it must be updated. The policy, or policies, should be informed by all the conditions requiring informed consent (e.g., assessment, treatment, specific interventions, significant modifications to interventions, sharing or exchanging information) that are indicated by regulatory, certifying, and other bodies like institutional review boards.

Along with a policy, there should be high-quality training in place that uses BST, appropriate consent documents and spaces to indicate consent on things like treatment reports and intervention plans, and an audit process for evaluating how well the policy is followed.

A helpful antecedent strategy to facilitate ethical behavior related to informed consent is to shift how you think and talk about it. Most of us were trained to "*get* informed consent." That's not wrong, but in our opinion, framing it as "giving someone the opportunity to *give* informed consent" might increase approaching the task in a collaborative manner that acknowledges and places the power singularly with the person being asked for consent.

Antecedent strategies related to ethical behavior for assent and assent withdrawal are essentially the same. It can also be helpful to collect and track data on things like client affect in therapy, frequency and duration of interfering client behavior that appears to be related to escaping the teaching environment, and the use of choice procedures.

Red Flags—Missing some or all the things listed in the antecedent section should function as red flags that consent and assent procedures and behaviors do not meet ethical standards and harm could be occurring. Framing informed consent procedures as technicalities or something that team members must go *get* can be red flags indicating an opportunity for improvement. Another red flag is if you, or others in the workplace, are unable to fluently list when and how informed consent must be obtained and/or you don't know where the appropriate forms or sections in documents are for indicating informed consent (or lack thereof in some circumstances).

Addressing—If you find that policies, resources, and training are missing or lacking in quality, well, that calls for a direct and collaborative conversation with those in a position to create or improve the needed things. You could say something like, "What do you think about our X (e.g., policies, training, forms, audit processes) related to consent/assent? I was thinking it could be good to review them against our ethics requirements and other recent scholarly work. What do you think? Could we get a workgroup together?"

If you are worried or aware that individuals are failing to get required informed consent, it's always best to have an open, collaborative, and compassionate conversation with the person to get the full picture. It's best to approach the conversation assuming ignorance or that you are missing something to increase the likelihood of a respectful and productive conversation.

Weekly Actions

MONDAY—THINK ABOUT IT

- Spend some time thinking about the training you did (or did not) get around informed consent and assent in graduate school, during your supervised fieldwork experience, and in the workplace. Can you identify any possible deficiencies that might invite unethical behavior or risk of harm to consumers and clients?
- Think about your most recent behavior related to consent and assent. As you reflect on that, what things can you identify that indicate ethical behavior? Are there things that you might im-

prove on or add to align your behavior with the ethics principle of "do no harm"?

- As you think about informed consent procedures, can you think back to a time when you or someone else may have failed to get informed consent when you should have? How did it turn out? Reflecting back, can you identify any actual or possible harms that resulted? Can you imagine a different approach that could have been taken? What would that have looked like?
- If someone asked you right now "When do you have to provide a caregiver the opportunity to give informed consent?" what would you say? Could you fluently describe the full range of instances that require informed consent and how to document it?
- Reflect on your thoughts and practices related to assent and assent withdrawal. Did you receive training in identifying and measuring assent withdrawal throughout a therapy or assessment session?
- Jot down notes and thoughts:

TUESDAY—LOOK FOR IT

- In your workplace, is there a policy focused on consent? How about assent? How about training and auditing for attaining informed consent, assent, and documentation for both?
- Can you quickly access any needed materials to support the informed consent process?
- Is there a process for what to do if informed consent was not obtained when it was required?
- Do your coworkers and others in the workplace talk about informed consent? What about assent? If they do, what kind of language do they use?
- What is the workplace culture like related to assent and assent withdrawal?
- Jot down examples:

WEDNESDAY—TALK ABOUT IT

- Begin talking about informed consent and assent with a colleague or a supervisor. Getting the conversation going will depend on the things you noticed in the workplace. Maybe you identified some red flags related to informed consent, in which case you might want to chat about opportunities for improvement. Maybe the informed consent practices are well aligned with our ethics obligations, so you might want to know how they were developed.

  Maybe there is no discussion or attention to assent, and you'd like to gauge interest in developing some assent and assent withdrawal processes based on recent scholarly work or propose some professional development on the topic.
- Consider talking to a trusted mentor or colleague/friend, especially if you know someone familiar with this topic. Ask them how they built their skill set. Ask if you can review any of their content or even observe a training, a meeting where a caregiver will be presented with the opportunity to provide consent, or a session that is assent focused.
- Chat about informed consent and assent with your supervisees or trainees. Maybe start by asking what they know about these topics. Then link the topics to ethics content and risks and benefits. If you have stories that came up during your Monday reflection, consider sharing them. You could have them read Breaux and Smith (2023) on consent and assent and then discuss.
- Jot down notes and thoughts from your conversations:

THURSDAY—ACT ON IT

- Now that you've taken stock of your language and behavior around your practices related to informed consent and assent, is there anything you want to change? If so, what, and why?
- What sorts of professional development activities might be beneficial to increase your knowledge and skills related to informed consent and assent?

- How about your supervisory and training practices? Did you identify any opportunities for improvement? If so, what, and how will you get started implementing those improvements?
- In looking at your workplace culture and practices, are there things that should be added or improved? Did you start talking to your supervisor or others about it? Can you map out an action plan?
- Jot down thoughts and tasks:

FRIDAY—REFLECT ON IT

- Spend some time reflecting on everything you have thought about, talked about, and acted on (or plan to) from this week. How are you feeling about everything?
- Reflect on how people reacted when you talked to them about these topics. How do you imagine they felt, based on their reactions and comments? How did their reactions make you feel?
- Review the tasks you jotted down, and develop a streamlined plan to address things over an appropriate timeline. Identify which tasks need to be prioritized based on risk of harm and available resources.
- Jot down something new you learned or something that surprised you:

Resources to Check Out

- To expand your understanding of barriers to informed consent and to get some great tips about how to improve your informed consent procedures, spend some time reading the "Quick Safety Issue 21: Informed Consent: More Than Getting a Signature" article by The Joint Commission (2022).
- The articles by Breaux and Smith (2023) and Flowers and Dawes (2023) provide useful definitions, behavioral conceptualizations,

and practical recommendations for taking an assent-based approach to service delivery.

- To learn more about consent and assent in relation to conducting research, check out the articles by Flowers and Dawes (2023), Mead Jasperse et al. (2023), and Morris and colleagues (2021).

Week 41

Check in—How are you feeling after spending a week tuning into your practices, and the practices of your workplace, related to informed consent and assent? We are guessing that you have identified some things that are going really well. We are guessing that you have identified some areas for improvement, right? And that is such a great place to be! We know that every week we are adding to your to-do list, which is rough. AND growth is such a beautiful thing.

So, hang in there. Take this week to review the scenario and get an organized plan for follow-up tasks. As always, this is a marathon, not a sprint. So, it's okay to spread your tasks over several months. Don't get too focused on the destination. This practice is about the journey. It is about the process of increased awareness and steady improvement.

Scenario—Review the scenario below, and then engage with the prompts. As always, the purpose is not just to give the scenario an ethics thumbs up or down, but to really reflect on the information provided and take a nuanced approach to considering areas that might be problematic or beneficial, and to really think about how you might react in a similar situation or how you might arrange your environment to avoid a similar ethical dilemma.

> *Rohan is a certified, licensed behavior analyst working in a clinic. About 6 months ago, Rohan completed an assessment and treatment plan, including a positive behavior intervention plan, for their client Vivian. The assessment process involved Rohan meeting with Vivian's caregivers to explain the assessment procedures, answer questions, and incorporate their ideas before asking them to sign an informed consent for assessment document.*

Once the assessment was completed, Rohan met Vivian's caregivers to review and explain the results, answer questions, and collaborate on a treatment and intervention plan. Once the plan was completed, including feedback and ideas from the caregivers, Rohan reviewed everything one more time and provided the opportunity for the caregivers to sign the actual treatment plan, indicating their informed consent to move forward.

Vivian's progress was excellent across all programming until about 1 month ago. The intervention plan for interfering behavior was based solely on positive reinforcement and did not include any targeted function-matched behaviors for increase because the treatment team felt that she had already acquired the relevant behaviors. Vivian recently demonstrated an increase in some interfering behavior that is making it challenging for her to engage with and complete work tasks.

Rohan consulted with Vivian's teacher because there is a current exchange of information consent and Rohan wanted to know if the teacher was also seeing an increase in these interfering behaviors. The teacher informed Rohan that they had seen an increase, so the teacher implemented a response cost procedure a week ago. Rohan, not wanting to waste any of Vivian's precious instructional time, immediately drafted a response cost procedure to be included in Vivian's intervention plan, trained the team, and made a note to review the modification with the caregivers in their next scheduled collaboration meeting in 2 weeks.

Jasper, another behavior analyst in the company, overheard the team meeting where Rohan introduced the new intervention component. He was concerned that Rohan might not have gotten informed consent, which Jasper thinks is necessary in this case because adding in a punishment-based intervention would likely be considered a substantial change to the intervention. However, Jasper knows that Rohan really understands informed consent. In fact, Rohan was the one who trained Jasper on the policy that says before assessments and interventions, you've got to get informed consent! So, Jasper figures that maybe he misunderstood or is misinterpreting the situation and decides not to say anything.

- **VALUES CHECK**—After reading the scenario, can you identify any core values that might be informing Rohan's or Jasper's behavior? Which of your values seem most related to this scenario?
- **IDENTIFY RISKS**—Circle language indicating a possible risk that was present, and then write down the risks/resulting harms.

  Can you tell that Rohan identified the same risks as you, or any risks at all? What about Jasper? Did he identify any risks? If so, what risks do you think he was attending to?

Be sure to consider specific aspects of the scope of risks or harms!

- **FIND & IMAGINE ANTECEDENT STRATEGIES**—Go back through the scenario, and draw an arrow to indicate anything you think functioned as an antecedent strategy or places where an antecedent strategy could have helped. Make a list of any antecedent strategies from the scenario, and evaluate if they were effective at preventing things like risks or actual harms. How do you know? Could anything have been improved or added? Now make a list of the antecedent strategies that you think could/should have been implemented, and briefly describe what it would look like to implement them and how you would know if they were effective.
- **IDENTIFY RED FLAGS**—Underline any red flags that you find, even if it seems that Rohan or Jasper did not notice them. For any red flags that occurred, write down what factors may have contributed to their ethical blindness (e.g., time constraints, lack of knowledge, authority, common practice, perceiving that there was no risk, lack of oversight). Can you think of any others?
- **IMAGINE INSTANCES OF SOMEONE ADDRESSING THE ISSUE**—Put an * next to a place where you think someone in the scenario could have taken action to address the issue. Assuming they did not, why? What were the barriers? What do you think should have happened? Describe the action that should be taken and how they would know if it was successful or not.

- **CONNECT**—Have you ever been in a similar situation or known someone who was? What was the outcome? What would you do differently, or recommend the other person do differently, now?

____

____

____

____

____

Taking Care of Business—Go back over the work you did last week, paying attention to tasks you identified for yourself or maybe for your workplace and/or your supervisory practices. After reflecting and completing the scenario activity, do you have anything to add? Maybe you want to spend some time reviewing any policies or training materials focused on informed consent and assent. Maybe you want to develop an audit process or develop or improve the tracking system for obtaining and updating informed consent. Take some time to prioritize your tasks, focusing on those that address risks to vulnerable individuals. Map your marathon. In fact, you might think of your tasks on these topics as more of a relay race, where you work collaboratively with colleagues to make sustained improvements over time.

Words of Inspiration—You know that your clients are the most important individuals in your work. Taking time to ensure that you have well-developed skills related to consent and assent will not only increase their safety and autonomy but will also likely strengthen the therapeutic relationship and increase collaboration. So, if you are nailing it, high fives to you! And if you find that you have some areas that are susceptible to improvement, well, welcome to our world and lean into this growth opportunity. You are doing great!

Weeks 42 & 43—Resting & Reconnecting

Week 42—Time to Rest & Take Care of Yourself

This is your last rest week, and we should probably let you in on a secret. We've been telling you that the focus of this book is to build your fluency with daily ethics practice. And that is true. But we have also taken a purposeful and committed approach to helping you lean into the idea that your health and wellness are important and that taking care of yourself is not selfish.

Hopefully, at this point, you have embedded practices to tend to your wellness and self-care into your daily and weekly activities. If so, you know what to do! If you are still struggling with making time for these critical practices, then maybe reconnect with the content in Chapter 3, Week 9, and other resources (e.g., articles, books, podcasts, presentations, and workshops) that can help you build this repertoire, because you are so valued and needed.

Monday–Sunday Self-Care & Wellness Intentions

You know the drill.

Monday Self-Care & Wellness Intentions

____

____

Tuesday Self-Care & Wellness Intentions

____

____

Wednesday Self-Care & Wellness Intentions

____

____

Thursday Self-Care & Wellness Intentions

Friday Self-Care & Wellness Intentions

Saturday Self-Care & Wellness Intentions

Sunday Self-Care & Wellness Intentions

Ideas for Regularly Scheduled, Recurring Self-Care & Wellness Practices

What do you want to integrate into your ongoing self-care and wellness practices?

Week 43—Time to Reconnect With Your Daily Practice

Deep breath. You've been working at developing your daily ethics practice for quite a while now, and hopefully things are feeling easier. Hopefully even the tricky topics and dilemmas are at least feeling less scary. Typically, the "Time to Reconnect" week is for playing catch-up. And you can totally use this week for that. We also invite you to use this week to go back over tasks you have completed and see if any need refinements or updates.

Getting new tasks done is important, for sure. It is also important to review tasks you have closed out with fresh eyes. So, you'll find the typical prompts that you've seen in the past for catching up. We also threw in some prompts to support you doing a review of tasks if that feels valuable to you at this point.

Review & Reflect— Check in with yourself to evaluate your progress toward building a fluent daily practice of focusing on ethics.

Celebrate— Go ahead, list those accomplishments, and praise yourself!

Refine, Plan, & Act— Review your list, prioritize, and get moving! Any lingering topics or tasks that you need to check in with?

Break out that list! Lean on your people (supervisor, mentor, colleagues) to help you keep taking action and making change for yourself, your organization, and your clients.

Tasks for This Week:

Tasks to Find a Cozy Spot on Your Calendar:

Look Ahead— You are coming to the end of the book's content, so check out the last few topics.

Tips for Reviewing Completed Tasks—Here are a few strategies that you can use to help review and potentially refine tasks you've completed.

- If you have created or revised training content, you can gather some social acceptability data. If there are people who experienced the original and revised versions, ask for their feedback about the changes. Ask folx who have completed the training if they feel more knowledgeable, skilled, or confident after the training as compared to before.

  Better yet, build a pre- and post-survey to accompany the training. If the training is competency-based (and most should be), how are people doing? Do the data indicate that the training is successful? If not, or if you aren't measuring mastery, take some time to make improvements or build in measurement processes and tools. Are you measuring maintenance of skills? If not, make plans to build that process.
- If you have created or revised policies, go back and review them with fresh eyes. Maybe even review them with someone else. Can you identify any improvements that need to be made? Is there a set schedule for reviewing and updating the policies? Do team members know about the policies? Have they been trained to implement them? Are you regularly auditing if the policies are being implemented correctly? If so, how do the data look? Is there an indication that you need to refine training or supports to improve implementation? If you don't conduct audits, build out that process.
- If you have added topics or activities to your supervision, take some time to evaluate if they are effective and if there is good social acceptability.

Inspiration to Reconnect— You may not be aware of it, but you are planting seeds now that will provide shelter and shade for future behavior analysts. As you continue with your practice, know that the effort you put in now will pay dividends in the future for you, your clients, your supervisees and trainees, and the profession.

Weeks 44 & 45—Discharging or Discontinuing Clients Professionally

Week 44

Check in—Hey you! Look at all the great content that you have absorbed so far! We hope that you took the last 2 weeks to recharge and let all the good stuff marinate to inform your daily ethical practice. As you prepare for this section, think back to Weeks 36 & 37 when we focused on transitioning clients proactively. Be on the lookout for commonalities and approaches that may also apply in this section.

As you self-reflect, also be aware of tasks or areas that are more challenging for you to connect with or tackle. What will you do to set yourself up to be most successful? Maybe you will set a timer, work with a colleague for accountability, or manage contingencies with your favorite activity or coffee after reading. Most importantly, you are here and doing amazing things!

Focus for the Week— When you were a kid learning to swim, did you use a life vest? You know the kind of life vest that has a bunch of foam floaties in it to keep you above the water, even if you don't want to be? As you got a little better and stronger in the water, your caregiver could take some of the foam pieces out so that you could feel your weight a bit more, and you started to rely on your strength and developed skills. Learning to swim was a lot of work and it was exhausting, but it was also exhilarating and you were getting stronger! When you were ready, they took them all out and you were swimming! Independent, strong, and successful.

Our role as service providers is not to give a life vest to a client and their family to use for the duration of their lives. We provide a life vest for as long as it is needed and partner with them to provide resources, tools, and support to strengthen their independence and abilities, all the while actively working toward a supported discharge from ABA services. Basically, we want to work ourselves out of each client's life. A key component in our ability to do this is to know where we are, where we are going, and how we are going to get there.

Where are we, where are we going, and how are we going to get there?

When you start school, you generally know there is a curriculum in place to get you to graduation after your senior year. Why would our services be any different? We start with the end in mind and partner with families to confirm exactly what they envision that end of services to look like. And we revisit these topics regularly throughout the service relationship. We don't want to aimlessly provide services and work through assessments and checkboxes without any real clarity and communication as to how and when it will end. To effectively do this, we need to encourage active participation and partnership with caregivers from the beginning and communicate, educate, listen, and coordinate care aligned with the plan that we create. As we discussed in Weeks 5 & 6, we need to talk about the last day of service just as much as the first day.

There are many ways a family can end services; some are better than others. The best possible scenario is a graduation from services, which includes a celebration! We refer to this as a discharge from services when we have worked with the family and client in tandem through their ABA lifecycle to be in a place where they no longer need the intensity of our services! This is the best-case scenario, what we all strive for, and serves as a strong reinforcer for many of us.

We will categorize any other reason for ending services as discontinuation from care. Perhaps the client is no longer benefiting from services. In some cases, services could even be contraindicated. What documentation do we have of the troubleshooting and efforts that we have engaged in to minimize barriers and support success? Why is the client no longer benefiting from care, and are we in agreement with the care team, including other medical professionals and the caregiver?

Did we do everything in our control to resolve barriers and support a successful outcome? Perhaps the reason for discontinuation from services are dangerous or harmful situations in the home (e.g., problematic alcohol or drug use, abusive or unsafe behavior by individuals in the home) that negatively impact therapists and that cannot be reasonably resolved. Maybe the client is discontinuing from services because they are moving out of the area. Maybe the caregivers are dissatisfied with services. Do you know why? Or maybe the funding source that made access to care possible is no longer available.

There are many roads that lead to the end of services, but remember that each road is unique to the family experiencing it, and it is the only road they are on. How will you use your experience, knowledge, and preparation to guide them along the path that is best, remove barriers, and support not just graduation from services, but the beginning of the rest of their life?

Values Check—What is your history and training when it comes to appropriate discharge or discontinuation of services? What do you believe to be the ultimate benefit of services, and how do your values match and align with this? You likely believe in access to care, so maybe discontinuing access to care feels in conflict with your values. Do you have clear guidelines as to when the need for access to care changes? This can be a difficult crossroads to be aware of and determine, let alone guide a family through.

Engage in some self-reflection around this area to determine where you stand and what potential weaknesses could exist in your standard approach. Is graduation from services something that is discussed at the end of treatment, or does it start from the beginning, even though that might make you feel uncomfortable? Do you value proactive planning and communication? Do you engage in perspective-taking and compassion as to why a family would or would not want to discharge from ABA services? How does your value of "clients first" conflict with loss of insurance coverage and discharge from services?

Take some time to think through these questions, because this lens will inform how you experience the situation and what you do about it. Be aware of your values, if and how you live them, and how they influence what you prioritize and spend time on, specific to supporting families through the discharge or discontinuation process.

Risks & Benefits—The risks of not having a thoughtful and effective graduation plan are significant and lasting. A hasty or poorly executed discontinuation from care could result in lost time for the client if another provider is not identified. Failing to discharge or discontinue a client without proper documentation of their programming could even result in the next service implementing interventions that you knew were ineffective or even harmful. Worst case, ineffective discontinuation of care could damage the caregivers' perceptions of ABA services to the degree that they no longer want services for their child.

The benefits of appropriate discharge and discontinuation of services ultimately mean that we did a great job setting our client and family up to be successful and increase independence in their future trajectory, or that we executed the hand-off to another professional (e.g., another behavior analyst, a teacher) in a way that placed the client first and minimized any negative impacts.

In the case of successful discharge from services because our services were no longer needed, a planful discharge decreases the likelihood they will need to return to care due to lack of thoughtful planning. We provide the right resources and training to appropriately support change agents in the environment who generalize and maintain learned skills while developing lasting learner repertoires.

Antecedent Strategies—When should you talk about discharge and discontinuation from services? The answer is, "Yes. Now. Always." The best time is always now and from the beginning. As discussed, we start with the end in mind. From the first meeting, we should discuss the end goal in coordination with caregivers and work backward to structure the environment in a way that will get us there. Have you explained to the caregivers that ABA is intended to be a temporary service to support skill acquisition and related independence, with the intention of the client developing repertoires to effectively learn from their environment with natural change agents, and not us?

The service agreement, which is thoroughly reviewed with caregivers at the outset of services, should clearly describe the events that could lead to discontinuation of services (e.g., chronic absences, unresolved conflicts, or persistent dangerous situations in the home) and what will be done to hopefully prevent discontinuation. It should also outline the steps for discontinuation by either party and the steps for discharge from services. The initial treatment plan and ongoing reauthorization reports should clearly outline the conditions under which discharge from services should and will occur. This includes the plan to titrate the intensity of services, and support training for caregivers and natural change agents to support the generalization and maintenance of learned skills in the absence of clinically trained individuals.

Okay, so we can start strong with a good plan, but the tricky part is staying on track and course correcting when we drift from the plan. This is inevitable, again, because we are human. So, good antecedent strategies are critical to ensure that our environment is structured in a way that minimizes drift and supports quick course correction. The key strategy here is to check ourselves against the plan. All. The. Time.

During your caregiver meetings, what are you talking about, and how are you spending that valuable time? Is your dialogue supporting coordination of care and the plan to discharge from services? For example, is there a spot on the standing agenda for your caregiver collaboration meeting to check in on discharge? You might say something like, "I know we are only 8 months into services, and we have a lot of important goals still to work on. But you know I am going to just touch base on eventual discharge, because that is our ultimate end goal."

When you review programs and progress and analyze data, are you making decisions and prioritizing treatment to support discharge from services? Have you completed assessments to inform the next treatment reauthorization in a way that supports discharge from services? Are your meetings, conversations, and focus intentional?

> *Intention is a combination of commitment and direction and is a critical consideration in effective care management. We have a critical window of service delivery where intentionality of care is paramount and time is of the essence, as it is one of the most precious resources that we have. Ensure that your behavior is intentional in coordination with caregivers to facilitate a successful partnership and discharge from services.*

There are a few important considerations that, if we pay close attention, can help us fulfill our ethical responsibility to facilitate discharge and termination from services. We have a responsibility to make **reasonable** and **appropriate** efforts to resolve barriers prior to considering discharge from care.

Check yourself against the following questions to ensure that your efforts are reasonable and appropriate:

- **The client is not benefiting from services**—*Before determining that discontinuing services is the most appropriate course of action, have you exhausted reasonable efforts to determine why your treatment plan is not producing the desired results? Has there been coordination of care with other providers? Have you consulted with your supervisor? Have you accessed other available resources to troubleshoot and explore different recommendations for assessment and treatment?*
- **There is a harmful situation that cannot be reasonably resolved**—*This may include all sorts of scenarios, as mentioned above. Has the clinical team had a clear, direct, and compassionate discussion with the family focused on facilitating a resolution? Can services be moved to a different environment? Can a different responsible adult be present throughout the duration of care? Are there different team members to support care delivery? Is additional training appropriate? Could the consent for* services and expectations of care be reviewed and re-signed?
- **The client requested discontinuation of services**—*There may be very appropriate reasons for a family to request discontinuation of services. Perhaps they are moving out of the area, in which case you should likely offer resources and care coordination with their new provider. Maybe they are traveling or managing through significant life circumstances and have decided to temporarily suspend ABA services. Are there alternative services such as consult or family training that could still be clinically appropriate in place of the full recommendation for the duration of the break?*

  If they are discontinuing services because of dissatisfaction, have you taken a magnifying glass to the health of the caregiver relationship and identified and addressed anything you could have or should be doing different with this family and all families? Have you solicited feedback or asked them to complete a patient satisfaction survey?

- **The caregiver is not compliant with ABA interventions, despite appropriate efforts to resolve**—*Caregivers failing to implement the treatment plan or who regularly miss or cancel sessions are telling us something. Maybe our training was insufficient, and they are doing their best to implement things. Perhaps we have not built a compassionate, collaborative relationship, and they are uncomfortable telling us that they are not on board with some of the interventions. Maybe there are life factors that cannot be resolved that make following the plan and sticking to the schedule difficult.*

  In any event, we should take a behavior-analytic approach to identifying the environmental contributors and work to address them, documenting the steps that we take. This may be an opportunity to evaluate the appropriateness of continuing services—after attempts to meet and review progress and successes—and review potential alternatives.
- **Services are no longer funded**—*This one is so tough. As a behavior analyst, we don't want something logistic like this to get in the way of medically necessary care, but insurance funding is one of the things that makes services possible. If services are disrupted due to funding, are there alternate resources or advocates who you can connect the family with? Are there alternate modalities of service delivery that can be provided to thoughtfully transition from more intensive services? How well does your organization proactively confirm benefits and eligibility to minimize an abrupt impact on services delivery? At the very least, can you ensure that the caregivers are well trained to continue implementing strategies?*

Red Flags—If you have no plan for discharge or discontinuation from services for any or all the clients in your care, you are waving a huge red flag in the winds of trouble. If the intake and assessment resources that you use do not prompt productive discussion related to the lifespan of ABA services and the importance of starting with the end in mind, that is a red flag. This will almost always place you in a position of reactivity. You are no longer controlling the direction and intentionality of services to support a graduation celebration, you are waiting for something else to come up and drive the discharge or discontinuation from services, which is not always positive.

If you have the resources and know what you should be doing but are not engaging in relevant conversations, YOU are a red flag! And that's okay! You have this week and next to reflect and build some strategies to better address this important topic.

There are other red flags that you might see around you, especially when we have not done a great job proactively planning, coordinating, and communicating. Have you or someone else ever said, "They just discontinued services and are moving out of state next week; it came as such a surprise." Don't get us wrong, things come up and happen, but it is less common to have a family just move out of nowhere. This is an opportunity to reread Weeks 5 & 6 and 21 & 22 to ensure that you are doing your part to maintain open lines of communication and support service delivery for those who need you. This is potentially a sign that you are less connected than you think you are and the family cares less about services than you think they do. Check yourself!

Again, surprises are often red flags, like your organization losing its staffing and capacity to coordinate effective care coordination, the family losing their insurance funding, or perhaps the current environment in which services are occurring is no longer supporting effective therapeutic services. These are all indicators that we need to be aware of and see from a distance, so that we can inform and plan a thoughtful path forward, instead of just reacting to the things that we could have seen coming. It is like you are driving down a road and keep seeing warning signs that the road is ending but ignore them and keep going. Yikes! Pay attention to the signs that access to care is threatened, and an appropriate and thoughtful discharge plan may not be possible, to best support your client and family.

Addressing—The best possible outcome is that we have a clear plan for discharge and graduation from services, we have a plan for addressing conditions that might lead to discontinuation of services, we have frequent check-ins with caregivers to track progress to our goal, and we review the discharge plan that is documented in detail. When the time comes, we celebrate! Or, in the case of us needing to discontinue services, we have a structured approach to ensure that we are placing the client first. If your actions and the structure of your organization do not support this outcome, then you might have some work to do.

One consideration is that we are better as a team than alone, especially in the context of decision-making. If your organization does not have a collaborative structure to formally review clinical progress, outcomes, and the recommendation for discharge, we recommend that you ask for one! Work with your supervisor and/or colleagues to put together grand rounds (yes, like you see in TV doctor shows), or perhaps a committee that formalizes the discharge review and approval process to ensure alignment to best-practice recommendations and considerations.

Review all that is in your control, in terms of what you communicate and how you prepare each family you work with, to start with the end in mind. How can we coordinate to get families excited for the outcome of discharge, and how they can be active participants in that?

Weekly Actions

MONDAY—THINK ABOUT IT

- Think about how you currently prioritize the discussion of discharge and discontinuation of services with clients and families. How does talking about discharge make you feel? Uncomfortable or a bit nervous?
- Have you ever had a family surprise you with a discontinuation from services?
- How do your values align or misalign with the importance of facilitating a thoughtful discharge from services plan?
- Reflect on the current clients in your care and how effectively you have started with the end in mind.
- Jot down notes and thoughts. Frame these thoughts in terms of where you are and where you want to be and go with no judgment. Show up better every day, and start today.

TUESDAY—LOOK FOR IT

- How often do you or others engage in discussions with caregivers about titration and discharge from services?
- When you or others discuss client discharge, at what point in the lifespan of ABA services does it occur? Toward the beginning, or the end, or throughout?
- Look at the resources, family paperwork, treatment plan structure, and other opportunities to clearly communicate and document the plan for appropriate discharge from services. Does the service agreement include information about discharge and discontinuation?
- Jot down notes and thoughts:

WEDNESDAY—TALK ABOUT IT

- Talk to a colleague, mentor, or supervisor about your reflections. Perhaps say something like, "I know the importance of thoughtful planning and communication to discharge from services, but this is not something that I intentionally do from the beginning or throughout service delivery. What are your thoughts on how we can more effectively structure the environment and resources to support a better plan?"
- Speak with your supervisor about what you think is going very well and what opportunities exist.
- Meet with your expanded clinical team and/or supervisees to engage in perspective-taking based on the current structure of your process and expectations.
- Spend some time talking to your supervisees and trainees about the importance of talking about discharge from day one, and why it is also important to outline the conditions that might lead to discontinuation from the outset. Talk about what structure is or could be put in place as a failsafe to inappropriate patient discharges, such as clinical rounds or a committee.
- Jot down notes and thoughts:

THURSDAY—ACT ON IT

- Perhaps you will review the plan and recommendations for your current clients. Are the programs and focus supporting the direction of care and anticipated timeline to titrate and discharge from services?
- Once you confirm the recommendations, review where this is documented. Is it outlined in a clear plan, and has it been communicated to the family as a partner? If not, this is your next step.
- What is your plan moving forward to ensure that future clients have the plan and work the plan with you from the very beginning? What questions will you ask, how will your treatment plan be structured, and how will your agendas for caregiver meetings focus in on where you are and where you are going based on progress and alignment to meeting goals? Maybe the service agreement needs some attention. Perhaps you can develop some scripts to support you having these conversations and training your supervisees and trainees.
- Jot down notes and thoughts:

FRIDAY—REFLECT ON IT

- Over the course of this week, we talked a lot about the importance of effective discharge planning, communication, and alignment to minimize risk to ourselves, our clients, and families. What was your biggest takeaway or aha moment? What is something that you do very well, or you see someone else do very well and aspire to?
- As you read through the reasons for discharge from services, how effectively do you make reasonable attempts to resolve barriers to access care and continue with services until a graduation and celebration can take place? Can you identify past situations where, in looking back, maybe you would do things differently now? That is okay; don't be too hard on yourself. Now is the time to identify the barriers that prevented you from taking a different path (e.g., lack of training or resources, feeling scared to address a concern) and how you can lean into making improvements.

- Jot down something new you learned or something that surprised you:

Resources to Check Out

- Be sure to review the *Continuity of Services* toolkit (BACB, 2021b). It contains resources for transitions, discharge, and discontinuation of services. This document will help you create and execute your plan to effectively support clients and their families and mitigate any vulnerability to yourself as a practitioner.
- The article by Green and colleagues (2023) presents the results of a survey conducted to assess perceptions and practices related to termination of services. The authors provide a host of valuable practice recommendations related to staff training and creating a professional will (i.e., a document that clearly outlines what should happen in the event of the provider's death or inability to continue to serve the client). The authors describe recommendations prior to, throughout, and near the end of treatment, and they include a handy table that outlines specific ideas across treatment phases.

Week 45

Check in—We reviewed a lot last week, and it is nearly impossible that it did not hit close to home. The relationships and services that we establish come to an end eventually, and we have a key responsibility in how that end is structured and experienced. As you have worked through some of the weekly items, or even just reflected on how well you have supported discharge from services throughout your career, think about how you will put this renewed insight and focus into action for the rest of your career.

Never underestimate the power and positive impact you can have in the lives of those you serve, but also those who observe and model your behavior.

Scenario—Read the scenario below, again, not just to classify what is ethical or unethical, but to view it with a lens of risk mitigation. How can we show up in a way that facilitates an intentional and positive discharge from services?

> *Apollo is a behavior analyst who has been practicing for 3 years and whose values align to clients first, but he has not had great training on thoughtful discharge and discontinuation of services, and the organization he works for does not have a structured way to present and plan with the end of services in mind.*
>
> *Apollo completes an intake assessment for a new client, Rowan, and his mother, Oakley. The team generally reviews what to expect from services but does not discuss discharge other than a statement in the treatment plan that is glossed over. At the 6-month review period, Apollo reviews the incredible progress that Rowan is making, but not in the context of his path toward graduation from services. In other words, he discusses the programs that Rowan is mastering, and the decreased gap of developmental milestones between him and his peers but does not discuss the reduction and discharge from services with Oakley.*
>
> *Another 6 months pass, and as the authorization ends, Apollo realizes that Rowan has met all the outlined goals and doesn't really need to continue with services. He calls Oakley and lets her know that instead of an authorization for continued access to care, he will be writing up a discharge plan and wants to plan Rowan's graduation party! Oakley is speechless. So much so that she reaches out to Rowan's supervisor, Elianna. Oakley shares with Elianna that services have been wonderful, and she doesn't disagree with the progress that Rowan has made, but she was taken aback at the abruptness with which discharge was suggested and the way it was messaged by Apollo.*
>
> *During their parent meetings, there were discussions about progress, but never about discharge, and she had no idea this was on the table. She does not feel ready or capable of supporting Rowan in the absence of ABA services. And although Apollo indicated that all*

programs were mastered, he did not ask Oakley if that appears to be the case at home and at school. Elianna recognizes Oakley's feelings and frustration, and although she is just as excited to see Rowan's progress, she reiterates the importance of their partnership and coordinated efforts to support a thoughtful titration of services within the next authorization period, with a clear focus to support a successful graduation from services in 6 months. Elianna indicates that she will review with Apollo and direct him to reach out directly to support next steps. Oakley expresses her gratitude for the collaborative dialogue and looks forward to discussing with Apollo.

Elianna meets with Apollo to gain additional insight, and Apollo says that he's been talking about discharge all along, because he assumed that as Rowan made progress mom knew they were getting closer to graduation from services. Elianna provides feedback that perception is reality, and there were missed opportunities to clearly outline and discuss how Rowan's progress got them further along the path to their end goal, which was discharge from services. Apollo and Elianna brainstorm how to modify the organization's treatment plan and clinical resources to ensure those conversations consistently happen with all clients and families.

- **VALUES CHECK**—How did you feel reading this scenario? Is one of your values effective communication and perspective-taking? Do you think that Apollo values those things too but did not have the resources and training to effectively execute in a way that he could live them? Sometimes it is not just a matter of living our values, but of having the training and awareness to effectively do so. Do you have the awareness, training, and resources to effectively prioritize what you indicate is important to you?

Be sure to consider specific aspects of the scope of risks or harms!

- **IDENTIFY RISKS**—Circle language indicating a risk was present, and then write down the risks/resulting harms.

  Did Apollo do anything to minimize the risk?
- **IMAGINE ANTECEDENT STRATEGIES**—What antecedent strategies should have been in place at different points in the service deliv-

ery journey, from the onset of services, during caregiver meetings, and at subsequent treatment reauthorization reviews? How could these antecedent strategies have impacted the outcome and experience?

- **IMAGINE RED FLAGS**—Underline any red flags that you see, even if Apollo didn't notice them. If you found a red flag (which you should have!), what other potential red flags should Apollo have on his radar?
- **IMAGINE INSTANCES OF SOMEONE ADDRESSING THE ISSUE**—Fortunately here, there was an opportunity to identify the communication breakdown. Put an * next to a place where you think Apollo could have taken better action. How did Elianna intervene to repair the relationship, inform the experience for Oakley and Rowan, and use that information to mitigate risk in the future?
- **CONNECT**—Have you ever been in a similar situation or observed someone who was? What was the outcome? What would you do differently, or recommend the other person do differently, now? Brainstorm and share your experiences and the other risks that were identified last week. Do you have the training and resources needed to effectively identify and proactively mitigate risk across all potential scenarios when discharging from services?

____

____

____

____

Taking Care of Business—Take a look back at your notes and highlights. What was profound and impactful for you? What are you most proud of that you are already doing or have done? Be sure you share and celebrate it with someone! Perhaps you didn't put anything into place, and that is okay too. Take a moment to ask yourself why and what you can do to increase the likelihood of putting what you learned into practice. If you try to do it all, you won't go anywhere at all.

So, have you taken action to reasonably and appropriately remove barriers to ensure that families can fully engage and benefit from care? If we were to give each of your families a call, do they know where they are in the lifespan of ABA services and the context of a graduation or discharge from services? Are they so excited about that? Have you thought about and shared any ideas with your supervisor to make it easier to have these conversations in a way that is positive and exciting? This is a tough topic for many, but your creativity, ideas, and authenticity will be a model and make us all better. Keep up the great work!

Words of Inspiration—When you live and engage with intention—the powerful combination of commitment and direction—the possibilities are boundless.

Weeks 46 & 47—Addressing Romantic or Sexual Multiple Relationships

Week 46

Check in—Holy guacamole! You're dang-near done with the book. How does that feel? We hope that your to-do list is not overwhelmingly long and that you are feeling more and more comfortable with your daily ethics practice.

Focus for the Week—You'll recall from Weeks 30 & 31, a multiple relationship happens when you, as a professional, take on one or more additional roles with someone you are responsible for professionally (e.g., client, caregiver, supervisee, trainee). Also remember that in Weeks 38 & 39, we got a bit more specific and addressed power differentials in relationships. Well, these 2 weeks are going to get even more specific and focus on multiple relationships that involve romantic or sexual relationships.

Virtually all professional codes of ethics in the helping professions forbid romantic or sexual relationships between professionals and their clients, supervisees, and trainees. Many extend that to the caregivers of clients. The risks of harm in this kind of boundary violation are clear and sometimes far-reaching, including things like damaging trust and producing feelings of confusion and guilt (Herlihy & Corey, 1992; Pope, 1988).

As a human, you can't control who you are attracted to, and you obviously can't control who finds you attractive. But you can control your behavior! If you're thinking *Yeah, okay. But I would never do something like this*, let us just suggest that to really build your skills related to identifying and addressing ethics-related issues, you need to imagine a world where it's possible that you or someone close to you could find themselves in this situation. So, let's spend the next 2 weeks leaning into this uncomfortable but critically important topic.

Note of caution: If you are someone who has experienced sexual assault, harassment, and/or related boundary crossing in the past, please keep yourself safe as you navigate this content. It may be helpful to cover these 2 weeks with a trusted colleague or friend who can support you and your self-care. If you happened upon this week at a time that makes interacting with this topic difficult for you, then just skip these 2 weeks and come back when you're ready. Or pause for a few days or weeks to get yourself in a headspace that will allow you to safely interact with the content for these 2 weeks. You know yourself best, so engage in self-compassion and self-evaluation and take good care of yourself.

Values Check—Of your values, which one or two jump out at you as being informative for this topic? Maybe all of them do. Spend some time reflecting on how you can use your values to minimize any avoidance responses that are probably at strength for you, and make space to call you into forward movement matched to your values.

Risks & Benefits—We already mentioned a few risks to the other person involved in a romantic or sexual multiple relationship with you, as a professional. Other risks include feelings of victimization, isolation, depression, and anger (Pope, 1988). In addition to risks related to mental health, both parties could experience damaged relationships with family members and coworkers.

For you, the risks include disciplinary actions from your employer and from any certifying or licensing body and possible legal action. "But the caregiver's child is no longer a client, or the supervisee was transferred to a different case," you say. Remember, the minute you start rationalizing, you know there's a risk. Are you confident that the client will never come back to your agency for services? Are you sure that you'll never have to step in to provide training, supervision, or performance evaluation for that team member? We never said we'd give you the answers to ethical dilemmas. We just promised that we would bring up challenging topics and difficult questions to support you on your journey of developing your skills for spotting and navigating ethics-related issues.

There is one primary benefit to avoiding multiple relationships that involve romantic or sexual interactions within your role as a professional: avoiding doing harm. You are in a caregiving and supervisory role that places you in a position of power over others, and you have an ethical and legal obligation to always behave in a way that places the needs of your clients, caregivers, supervisees, and trainees first. And *your* supervisor is in a similar role to you, so they also should take care to avoid creating multiple relationships with you.

Antecedent Strategies—Okay, sit down again. Maybe get in your comfy place. The best antecedent strategy to avoiding romantic or sexual relationships with people you are responsible for is to *truly believe* that you are at risk of this happening. Why? Well, the development of romantic or sexual attraction can be subtle and develop over time, and you can't control those thoughts and feelings. So, catching those thoughts and feelings early and taking action to protect your clients, caregivers, supervisees, and trainees is the key to keeping everyone safe.

Avoiding these types of multiple relationships is all about catching red flags and engaging in self-control. Knowing yourself well, and without judgment, is key here. Are you someone who is highly reinforced by attention and being in relationships? Are you someone who has a history of impulsive decisions, particularly related to getting involved in romantic or sexual relationships? Finding someone attractive is, in truth, harmless. The risk lies in failing to take objective steps to prevent yourself from acting on those thoughts or feelings and failing to notice AND acknowledge subtle changes in behavior that are influenced by them.

So, the first preventative strategy is for you to reflect on yourself and your past thoughts, feelings, and behavior in this area. The second preventative strategy is to make it a regular practice to honestly scan your thoughts, feelings, and behavior for and toward your clients, caregivers, supervisees, and trainees, and take action if needed. Finally, it's a good idea to have a plan of action for how you'll respond if you notice red flags in your relationships or the relationships of others. This topic is complicated, so it's best to develop some steps for addressing it before you find yourself smack in the middle of the need to act.

Red Flags—The red flags here are pretty obvious; the issue is failing to be honest and admitting to them. Here are a few we thought of:

- having romantic or sexual feelings or thoughts
- thinking about the person far more frequently than other similar people
- extending meeting time with the person

- looking at the person or photos or videos of the person more frequently than you do for other people
- finding excuses to be in proximity (especially in close physical proximity) to the person
- finding ways to be alone with the person
- contacting the person outside of the professional context

It may be clear to you when someone else is engaging in some of these behaviors, and obviously you should be paying attention to these red flags in others, particularly with your supervisees and trainees. But don't be fooled into thinking that you are immune—it's critically important to make space to believe that you are at risk of engaging in these red flag issues. And honestly, that's a good thing, because catching the red flags allows you to take action in service of your values and protect the other person and yourself.

Addressing—What should you do if you find yourself engaging in any of these red flags? Well, the first thing is to be kind to yourself. Remember, you can't control your thoughts and feelings, and by acknowledging that some red flags are showing up, you're taking the most important step to keeping everyone safe.

The second step is to take some committed action to prevent the possible multiple relationship from developing. That could involve writing down your thoughts and feelings, the risks of taking action on them, and strategies for engaging in behavior to keep everyone safe (e.g., never meeting alone, sticking to a strict agenda, refraining from engaging in physical touch). You might (and by might we mean should) consider talking to a trusted supervisor or mentor. Writing or speaking your thoughts and concerns combats the risks of engaging in rationalizing and increases the likelihood that you will take action to prevent things from progressing in a harmful and potentially dangerous way.

Depending on your history and how strong your feelings are, you may even consider transferring the client, supervisee, or trainee to a different behavior analyst. Whatever the case, addressing this requires honest self-reflection to move in a meaningful way toward keeping everyone safe.

In the unfortunate event that you did not attend to red flags and a romantic or sexual relationship has developed, you're obligated to immediately bring this to the attention of your supervisor and work to resolve the multiple relationship. How? Well, we can't answer that for you, as each situation is unique and nuanced. Resolving the multiple relationship could involve ending the romantic or sexual relationship, but likely the professional relationship has been changed in ways that can't be fixed. So, you might also have to transfer the individual to another professional.

Maybe you feel that you're deeply in love with the supervisee or caregiver. If that is the case, and you want to maintain the personal relationship, you'll need to resolve the professional relationship. The solution could involve transferring cases, or in some cases, the right thing might require you to leave the company. You might also have to self-report this ethics violation to relevant certifying or licensing bodies.

Maybe you're concerned that another behavior analyst is at risk of being in or is in a romantic or sexual multiple relationship. If so, this likely feels uncomfortable for you, which means you're at risk of avoiding having a discussion in favor of sparing their feelings. But avoiding bringing this up means that you're placing the other person in the relationship, and the behavior analyst, at risk of serious harmful consequences.

So, lean into this discomfort and prepare to have a difficult conversation. On the other hand, you might find yourself riding into this interaction on your noble steed of righteous indignation! Pull on those reins and slow yourself down. Approaching someone from a position of "How could you do such a thing?!" increases the chances of the conversation going poorly and is in opposition to our science. Instead, activate your compassion and approach the conversation by taking a circumstances view that requires you to acknowledge that behavior is a product of people's past and current environments.

Weekly Actions

MONDAY—THINK ABOUT IT

- How do your values support you in committing to make space for the fact that you're at risk of finding yourself on the road to a romantic or sexual relationship with a client, caregiver, supervisee, or trainee?
- What is your learning history related to romantic and sexual relationships? Are you someone who values this type of at-

tention from others? Have you engaged in "harmless flirting" in the past with a supervisee or trainee or caregiver that never "went anywhere"? Have you engaged in this type of multiple relationship in the past? No judgment here, just a gentle prompt to engage in honest self-reflection. We invite you to move away from guilt or shame here in service of your growth and ability to place the needs and safety of others in the professional context front and center.
- Think about how you have historically dealt with impulses and desires. Again, no judgment here. But you're the only person who has access to information about your past behavior and your thoughts and feelings related to this topic. Again, do your best to engage in self-compassion, and hold any thoughts or feelings of guilt or shame lightly. You are human and this work is hard. Give yourself grace.
- Jot down notes and thoughts:

TUESDAY—LOOK FOR IT

- Can you identify any examples of these types of multiple relationships that you've witnessed or heard of in the workplace? In graduate school? In other professional contexts? What was the outcome for all parties involved?
- Is training on this topic available or required at your current or former workplace? In graduate school?
- Are there policies in your workplace to help prevent these types of multiple relationships? If there are policies, do they allow for things like inter-staff dating, and if so, under what conditions?
- Jot down examples:

WEDNESDAY—TALK ABOUT IT

- Find time to chat about this topic with your colleagues or your supervisor. If your organization provides training or has a policy on this topic, you can use that as a conversation starter. You might ask, "I wonder if we could chat about the policy/training on romantic and sexual relationships? I'm interested in making sure that I'm leaning into this topic so I can watch for red flags and have a good plan of action if I happen to notice any."

- Find or create some scenarios that address this issue, and tackle them with your colleagues, supervisor, or mentor. If possible, ask your colleagues, supervisor, or mentor about their experiences with this topic. If you have trusting relationships, see if it feels comfortable to share examples of how either or both of you have addressed red flags or these types of multiple relationships in the past.
- Bring up this topic with your supervisees or trainees. Again, if your organization has a related policy or training, you can use that as an introduction to the topic. Talk about related ethical obligations, and use scenarios to walk them through how to take an analytic approach to looking for and addressing emerging or actual romantic or sexual multiple relationships. Be sure to talk about the fact that we are all at risk and must engage in regular self-checks to look for red flags we can attend to.
- Jot down notes and thoughts from your conversations:

THURSDAY—ACT ON IT

- Revisit your thoughts from Monday and conversations from Wednesday, and make a list of any risk factors for yourself. These might be related to your risks of engaging in these types of relationships based on your history, or risks related to avoiding addressing these types of relationships for yourself or with others. Consider writing down some red flags that might present themselves to you.
- Spend a little time making an action plan that you can trigger if you see any red flags. Maybe write down some strategies for addressing red flags (e.g., engaging in competing self-talk when you find yourself thinking about a caregiver or trainee in a romantic or sexual way—"It's reasonable that I find this person attractive, and I need to be careful about my behavior because I value this person's safety and growth and I have a professional role to place

their needs and safety first."). Identify and write down the names of a few trusted people you could go to if needed.
- Review any policies or training available in your organization. Are there areas for improvement? If so, is it possible to have a conversation with your supervisor about making some adjustments to the policy or training?
- Jot down thoughts and tasks:

FRIDAY—REFLECT ON IT

- Think back on this week. Did this topic make you feel uncomfortable? Reflect on your conversations from Wednesday—can you identify that the topic made others uncomfortable? How do you know?
- Reflect on the thoughts and feelings you had about this topic on Monday. Compare them to your thoughts and feelings today. Have they changed? How?
- Remember when we said that one of the most important things you can do related to this topic is to *truly believe* that you are at risk of this happening? Where are you on that journey? If you are feeling resistant to this idea, what do you think might be getting in the way of you accepting this possible risk?
- What will you do to ensure that you continually make space to evaluate your thoughts and feelings related to others in the work setting?
- Jot down something new you learned or something that surprised you:

Resources to Check Out

- If you want to dive in a little deeper to multiple relationships, psychologists' perceptions about them, and some of the harms caused by them, Afolabi (2015) has a nice article that reviews the literature and a decision model along with case examples.

- Consider checking the APBA toolkit addressing multiple relationships (APBA, n.d.) that we mentioned in Weeks 30 & 31. There's a comprehensive document to help you learn more about this topic, engage in self-evaluation, and take action. There's also an assessment and planning template for you to use, and some resources that might be useful for your organization.

Week 47

Check in—Woof. Last week we covered a pretty heavy topic—multiple relationships of the romantic or sexual variety. Take a moment to check in with yourself. How are you feeling? You doing okay? Let us just take a moment to say how proud we are that you are leaning into this topic and committing to working on ethics content regularly.

Scenario—Read the scenario, and complete the prompts. Keep in mind that romantic and sexual multiple relationships can come in a wide variety of forms.

> *Beatrice is a behavior technician who just started working at a provider agency a few months ago. She is working with four clients. Three of her clients are supervised by Patricia, who is a great behavior analyst. Beatrice gets along well with Patricia and loves learning from her. The fourth case is supervised by Rian, who is also super smart. In talking with Rian, she realizes how much they have in common, how easy it is to talk to Rian, and how much fun they have even though they are getting work done.*
>
> *Rian has been supervising for 2 years and loves the work they do. They enjoy working with Beatrice, who is catching on so fast and is really funny. Rian isn't even aware of it, but they look forward to overlapping with Beatrice, and they tend to put more effort into supporting Beatrice with her client and professional development. Rian regularly tries to reflect on how they are feeling toward clients, caregivers, supervisees, and trainees to watch for red flags.*

In reflecting on Beatrice, they can see several red flags. Rian finds themself looking at Beatrice a lot, even when they are not officially observing Beatrice. Rian also notices that they often try to find activities that allow them to interact with Beatrice and her client during sessions, like board games or games with physical components like tag. They have also been texting Beatrice more than is required for work, and they often use language that, upon reflection, seems flirtatious. Rian really likes Beatrice and would like the opportunity to spend more time with her and just see where things might go.

At the same time, Rian understands their ethical obligations related to multiple relationships and wants to make sure they do not violate any ethics standards. Rian asks their supervisor to reassign that client to Patricia, since Beatrice has three other clients under her, making the point that doing so will give Patricia more time to train and supervise Beatrice. Rian continues to supervise Beatrice and the case during the transition, and their attraction to Beatrice continues to grow. Rian can tell that Beatrice likes them too, as they steal glances at each other and share inside jokes during sessions. Once the case is fully transitioned, Rian reaches out to Beatrice to invite her on a date, and she agrees.

- **VALUES CHECK**—After reading the scenario, which of your values are most related? Can you identify any values that Rian might be struggling with?

- **IDENTIFY RISKS**—Circle language indicating a risk that was present, and then write down the risks/resulting harms.

  Did Rian or Beatrice also identify the same risks? Did they take action to minimize or avoid the risk? Are there other risks or actions you can think of that could be related to this situation?

Be sure to consider specific aspects of the scope of risks or harms!

- ***How big/bad is the risk/harm?***
- ***Who is/would be impacted?***
- ***Was the risk/harm possible, or did it actually occur?***
- ***Is it likely that the actual or possible risk/harm will occur again in the future?***

- **FIND ANTECEDENT STRATEGIES**—Go back through the scenario, and draw an arrow to indicate anything you think functioned as an antecedent strategy. Make a list of the antecedent strategies and who implemented them, and evaluate if they were effective at preventing things like risks or actual harms. How do you know? Could anything have been improved or added?
- **IDENTIFY RED FLAGS**—Underline any red flags that you find, even if Rian or Beatrice didn't notice them. If they didn't notice them, write down what factors may have contributed to their ethical blindness (e.g., time constraints, lack of knowledge, authority, common practice, perceiving that there was no risk, lack of oversight). Can you think of any others?
- **FIND INSTANCES OF SOMEONE ADDRESSING THE ISSUE**—Put an * next to any language indicating that Rian or anyone else took specific action to address the issue. Make a list of those actions, and then describe if they were effective or not, how you know, and anything you would have changed or added. Were there other points in the scenario where you think one of the individuals could have taken a specific action to address the situation? If so, what, what would that look like, and how would you know if it was successful?
- **CONNECT**—Have you ever been in a similar situation or known someone who was? What was the outcome? What would you do differently, or recommend the other person do differently, now?

Taking Care of Business—Use this week to revisit the tasks that you identified last week. A lot of the tasks related to last week are about engaging in honest and compassionate self-reflection, so take it easy. Don't push yourself if this topic makes you uncomfortable or brings up difficult thoughts and emotions. If you can, try to get them done this week. If not, identify at least one for completion this week, and then put the rest of the tasks on your calendar for future weeks. As we are getting close to the end of the book and your year of developing intentional ethical practice, you might find yourself scheduling these tasks for when you have completed working through the book, and that's okay! We've also got time planned for you to work on any remaining tasks during Week 51.

Word of Inspiration—You are doing a wonderful job leaning in to challenging and uncomfortable topics. As a supervisor and leader, you should be proud of yourself for making space to reflect on this topic and how you will work diligently to protect your clients, caregivers, supervisees, trainees, and yourself from the harms that can come from stepping over the line of a professional relationship into a romantic or sexual multiple relationship that could be harmful.

Weeks 48 & 49—Dealing With Interruptions, Transitions, & Terminations of Supervisory Relationships

Week 48

Check in—We're getting close to the end, folx! Hopefully working through the topics has been easier to implement into your daily activities over the past 47 weeks. Maybe you've even started thinking about how to continue keeping ethics in your daily practice beyond this year. More on that later! Okay—let's go ahead and jump into our last biweekly pair.

Focus for the Week—In previous biweekly pairs, we've discussed how to design and evaluate your supervision, how to ensure you take responsibility for your supervision practices, and how to ensure you are taking all the necessary steps to ensure continuity of services and appropriate discharge from services for clients. So, of course, it makes sense for us to spend some time covering content related to transitioning and terminating supervision with our supervisees and trainees.

And this might be another topic that, as we start to work through it, has you thinking: *Wait. I haven't really thought about this before!* or *Oh no—I don't think I really followed through on this in the past.* As we said before—give yourself some grace. The purpose of discussing these topics and putting them in your daily practice is to get you thinking about your practices now and to help you successfully assess and address these scenarios in the future.

There are many reasons why a supervision relationship might come to an end. For example, it may end naturally when a trainee finishes up their supervision hours. And that's a lovely reason for ending the supervisory relationship. Another reason might be that a supervisee is assigned to a different client that is not on your caseload. Or maybe you or your supervisee or trainee leaves the organization. One of you might get a promotion within the organization. Or maybe there was a violation of your supervision contract, and the supervisory relationship might need to be terminated. Maybe you have a client who needs additional support for a period of time, and you're unable to provide adequate supervision for a trainee or supervisee for that time frame.

And just as we put time and effort into transitioning or terminating services with a client and ensuring that we minimize lapses in services, we should put the same time and effort into transitioning or terminating the supervisory relationship and ensuring that we minimize gaps in supervision for our supervisees and trainees. There are three types of supervision-related events to consider: (1) interruptions to supervision, (2) transitioning supervision, and (3) terminating supervision. Sometimes we'll cover all three together, and sometimes we'll separate them when the considerations are distinct.

Values Check—As you're thinking about what values you have identified that most align with this topic, you might take a moment and look at your notes and see what values you selected when you were working through the other biweekly pairs that covered supervision topics, Weeks 7 & 8, 13 & 14, and 28 & 29, and the biweekly pairs that covered transition and termination of services for clients, Weeks 36 & 37 and 44 & 45. As you think about the values that align with this topic, are you identifying similar ones to those you identified as you worked through those past biweekly pairs? Have you identified any new ones that fit here that you didn't identify before? Keep these values in mind as you work through this content.

Risks & Benefits—Many of the risks are the same when supervisory activities are interrupted, transitioned, or terminated. All three of these changes in the supervisory relationship can negatively impact clients. A supervisor missing observation sessions could mean that technicians continue to implement programming incorrectly because they are not getting the necessary training and feedback. This can negatively impact client progress and can even result in physical harm if a client engages in dangerous behavior (as we already discussed in Weeks 36 & 37).

Changes in supervision, for any of the three reasons, could also prevent the supervisee or trainee from accruing or receiving the amount of supervision required by a licensure board, certification body, or insurance payor. This might have larger impacts if it occurs multiple times and could even delay or prevent the trainee or supervisee from obtaining or maintaining any certification or credentialing by any of those bodies. These unmanaged supervisory changes can also impact your supervisees' or trainees' perceptions of the support they are receiving, which may in turn chip away at the health of your supervisory relationship (e.g., they start to doubt your skills as a supervisor, they feel you don't value them).

Mismanaged transitions or terminations can negatively impact the supervisee's or trainee's progress if information about where they are in their supervision (e.g., areas they are currently receiving training on, previous feedback, areas where they have already demonstrated competence) is not shared with the next supervisor. These situations could result in wasted time if the new supervisor needs to conduct assessments or covers content that you already covered but failed to communicate.

And specific to terminations, if the reasons for possible termination of supervision are not clearly communicated at the outset, the supervisee might be caught off guard and might even feel that you are behaving in an unethical way, especially if they feel that the termination is unfair or unwarranted.

So, what are some benefits of preparing for and managing potential or actual interruptions in your supervisory activities? Yeah, you guessed it, the benefits are the opposite of the risks: (1) minimizing the potential negative impact on clients, (2) supervisees and trainees being able to meet the requirements and gaps in supervision being minimized, and (3) the next supervisor being able to take your information on the current strengths and weaknesses of the supervisee or trainee and use that information to help continue supervision as seamlessly as possible without wasting time.

Being up front and planful with any of the three types of changes to your supervisory relationships should also protect your relationships with your supervisees and trainees (i.e., they know you have a plan to make sure they are receiving the support they need). And finally, taking a proactive approach means you'll be meeting your ethical obligations as a supervisor and protecting yourself from a potentially sticky situation if a supervisee or trainee feels that you did not terminate supervision fairly, because you'll have documentation on the plan you followed.

Antecedent Strategies—There are a handful of active steps you can take to prepare yourself and your trainees and supervisees in case of an interruption, transition, or termination of supervision.

---

1. *Clearly communicate the reasons why a supervision relationship might be interrupted, transitioned, or terminated. You might proactively identify the specific reasons why you might terminate the supervisory relationship (e.g., egregious misconduct, not meeting expectations for X number of months following a performance management plan) and communicate these clearly at the start of the supervisory relationship. You might even think about including these things in a contract or other document (e.g., form, agreement) so it is, you know, documented. This is probably also a great time to discuss reasons why the supervisory relationship might be transitioned or interrupted and what the process is in all three of those situations.*
2. *Identify the steps that will be taken if an interruption, transition, or termination occurs. Include a plan for documenting the steps taken and how they will be communicated to the relevant parties (e.g., trainee or supervisee, next supervisor). You may consider including the duties of each party if one of those situations occurs in any relevant supervision contracts.*
3. *Ensure that all documentation is up to date, and share it with the necessary people (e.g., supervisee or trainee, next supervisor, organization). Also, make sure that everyone is aware of any organization, licensure board, or certification body requirements for retaining documentation.*
4. *Review any relevant supervision requirements, and ensure that your processes meet the requirements. Determine if the processes provide steps to prevent any lapses in supervision (e.g., a plan if a supervisor must take extended leave).*

Red Flags—These red flags are pretty similar to the ones we highlighted in Weeks 36 & 37 and Weeks 44 & 45. You might be at risk if you find yourself thinking something like *I don't have time to transition this supervisee.* Or *Hey, they are the reason why I am terminating the supervisory relationship—I shouldn't have to spend the time making sure their documentation meets the requirements.* Or maybe you're realizing that none of your supervisors had conversations with you about why the supervisory relationship might end and what the next steps would be after it does, so you're not quite sure how to approach that conversation. Another red flag is if you are depending on the organization you work for to manage the interruption, transition, or termination, rather than taking ownership yourself.

Addressing—Let's talk about some ideas for addressing these situations. Maybe you've identified that you don't have a plan for interruptions, transitions, or terminations of supervision. The first step you can take is creating one. You can reach out to your organization or supervisor to see if they have one you can use as a model. Or maybe you can collaborate with a colleague to create one. You can also review the "Resources to Check Out" section for a few sources to use.

Next, list out the steps that you'd take to address each of those situations. What are some barriers you might contact while in these situations? Can they be addressed with the antecedent strategies listed above? Are there any other strategies you can implement to address the barriers?

Did you also identify that you haven't addressed these situations with current supervisees or trainees? Well, now is the time to prepare for having these conversations, clarifying expectations, and discussing what will happen if there is an interruption, transition, or termination of the supervisory relationship. This also provides a model for your supervisees and trainees for their future as a supervisor.

Weekly Actions

MONDAY—THINK ABOUT IT

- Take a moment and reflect on your supervision experience as a supervisee or trainee. Was addressing interruptions, transitions, or termination of supervision discussed or modeled for you? Were you taught how to prepare yourself for these situations as a supervisor? Or, if you've been in any of these situations now as a supervisor, have you had to learn by trial and error?

- How do the values you identified above align with ensuring that you are planning for interruptions, transitions, or termination of supervision? How can you use these values to help you stay motivated to ensure that you are taking the necessary steps to address these situations when they occur?
- Spend some time to reflect on your experience as a supervisor. Can you identify times when you've been involved in managing an interruption, transition, or termination of supervision (e.g., as the supervisor experiencing the interruption, transition, or termination, as the behavior analyst taking over for another supervisor who has experienced an interruption, transition, or termination of supervision)?
- Can you identify risks or harms that occurred because you did not plan and/or effectively support the interruption, transition, or termination? Can you identify what variables (e.g., not enough time, damaged relationship, no policy or procedure in place) were occurring at the time that may have impacted your ability to appropriately address an interruption, transition, or termination of supervision?
- Take a moment and pause. The purpose of today's prompts was just to self-reflect.
- Jot down notes and thoughts:

TUESDAY—LOOK FOR IT

- Let's start with thinking about the organization you work for (or your organization if you have your own). Are there processes already in place for supporting your supervision, particularly, interruptions, transitions, or terminations? Is there a supervision contract your organization uses that lists out the reasons why a supervisory relationship might be terminated? Is there a plan for covering supervision if a behavior analyst is going to be out for a short time (e.g., PTO, illness, family emergency)? Think back to Weeks 7 & 8, 13 & 14, and 28 & 29, when you spent some time reflecting on developing and evaluating your supervisory practices. In your reflection during those weeks, did you identify that your organization commits to supporting high-quality supervision, and continuity of supervision at that?

- Are there any resources or supports that would be helpful to add or create to help avoid or address these situations (e.g., checklist, contract template)?
- Are the reasons why supervision may be interrupted, transitioned, or terminated clearly documented and communicated with all relevant parties (e.g., supervisees, trainees, supervisors)?
- Jot down examples:

WEDNESDAY—TALK ABOUT IT

- Now that you have had some time to reflect on your experience and make some observations in your organization, take some time to chat with a colleague or mentor about what you've learned. You might see if they are comfortable sharing their own experiences as supervisors and how they work toward having continuity in supervision. Do you have any similarities in your reflections and observations? What about differences?
- Schedule some time to talk to your supervisor about your reflections and observations. You can share with them the areas that you feel prepared to manage those situations in as they arise. Then, if you're comfortable, you can share areas you'd like their support in. Maybe you've also identified areas where the organization could better support supervisors with maintaining continuity of supervision. This might feel a little uncomfortable to bring up, but let's take a moment to recognize that by this point, you've probably had quite a few difficult conversations, and you likely have experienced some growth in those skills.

  So, if you need to identify some potential improvements and have a conversation with leadership in your organization about it, take some time to reflect on how your previous conversations have gone, make some notes on what you want to address and how you want to address it, practice it with a colleague (if that strategy is useful for you), and jump on in. Remember, you both are here with the goal of providing excellent supervision to support the profession, and this conversation will only help your organization improve.

- Did you note that you have not discussed the reason and plan for interruptions, transition, or termination of supervision with your supervisees and trainees? Well, now is the time to have that conversation with them, and document that you had it of course. And you know what? It's okay to acknowledge that you should have had this conversation at the start of the supervisory relationship. You are modeling that when you identify that you missed a step, you should address it as soon as you become aware of it. It might be a bit awkward or uncomfortable, but remember, you are modeling how to own and change your behavior.
- Jot down notes and thoughts from your conversations:

THURSDAY—ACT ON IT

- Formally review your process for addressing supervision interruptions, transition, and termination. If you or your organization don't have one, now is the time to start drafting one. Go back and review the antecedent strategies and red flags, and make sure they are included and addressed, respectively. It may be helpful to have a colleague, supervisor, or mentor review the process (especially if you have created one from scratch) and give you feedback.
- Create a list of reasons why supervision might be interrupted, transitioned, or terminated. Review the list with a colleague, mentor, and supervisor. Do they agree? Do they have any suggestions?
- As we wrap up the biweekly pairs specifically discussing supervision, you might identify that you have some leftover tasks from previous weeks. Now is a great time to go back through those lists and prioritize the tasks you have left and those that you've identified this week. You're not going to get it done all at once, so take the time to create a plan that will set you up for success.
- Jot down thoughts and tasks:

FRIDAY—REFLECT ON IT

- How are you feeling as you wrap up the week? Did the reflections and conversations seem to be easier now that you've worked through this process with similar topics, for example, when you were first thinking about how you approach conversations with caregivers and reflecting on if you were just going through the motions or actually listening to their experiences and feedback back in Weeks 5 & 6? The same? More difficult?
- How will ensuring that your supervisees and trainees experience continuity in their supervision impact them in the future? How do you think it will impact them as supervisors one day? What do you think would happen if you didn't work to ensure continuity for them?
- What will you do to ensure that your process for addressing these situations is effective?
- Jot down something new you learned or something that surprised you:

Resources to Check Out

- The *Continuity of Services* toolkit (BACB, 2021b) focuses primarily on interruptions, transition, and discharge of client services, but we think that the same considerations can be applied to thinking about the same situations with supervision. You'll just have to do some translating!
- The supervision and mentorship book by LeBlanc, Sellers, and Ala'i (2020) provides guidance on tackling problems in supervisory relationships, including addressing when they are ended.
- In their article, Turner and colleagues (2016) discuss considerations for terminating the supervisory relationship.
- Sellers, LeBlanc, and Valentino (2016) discuss barriers to supervision in their article, including what to do if you identify that a supervisory relationship must be terminated.

Week 49

Check in—How are you doing? Are you finding ways you can incorporate the actions that you identified above into the way you also transition or discharge a client from services? Those processes will likely interact, as you may find yourself transitioning a supervisee or trainee who works with a client when you transition the client to another behavior analyst in your organization. This might be another topic where you've identified that there were times when you did not address the situation in the best way possible. Again, reflecting on those past situations may not feel good, but remind yourself that now you can do better.

Scenario—It's time for your last scenario! Take time to review the scenario and review the prompts. Use this example to think about the risks, benefits, values, and red flags that might come into play in this or similar situations. You might even identify some actions you want to add to your list as you work through this scenario.

> *Della has been a trainee accruing fieldwork hours with Rochelle for the past year. Rochelle and Della have a supervision contract that clearly outlines the reasons why their supervisory relationship might be terminated and the expectations of both parties. Rochelle discussed this at length with Della when they reviewed the contract, and both parties signed the supervision contract.*
>
> *Last week, Rochelle had an emergency. She has not been back to work and has limited communication with the team. Della has missed one supervision meeting already and is worried that if Rochelle is unavailable for much longer, Della may lose a significant number of supervision hours. Because she did not want to be a burden to Rochelle during the emergency, Della did not reach out to Rochelle directly. She reviewed the supervision contract but did not find any information about the expectations in the event the supervisor is unavailable to provide supervision for some time due to a planned or unplanned event.*
>
> *When Della reached out to Anya, who was covering supervision for Rochelle's clients, Anya said that she could not cover Della's super-*

vision hours outside of the supervision required for the clients and was hesitant to sign a supervision contract with Della. When Della reached out to the clinic director, she was told that they did not have a timeline for Rochelle's return and Della would just have to be patient.

- **VALUES CHECK**—If you had to take a guess about Della's values, what might one or two of them be, and how can you tell based on the scenario? Do they align with your values? How about Rochelle's values? And Anya's?
- **IDENTIFY RISKS**—Circle language indicating a risk was present, and then write down the risk/resulting harms.

  Did Della identify potential risks? Did she take action to minimize or avoid the risk? Are there other risks or actions you can think of that could be related to this situation? What about Rochelle? Did she identify potential risks and take action to minimize or avoid the risks? If Anya had provided supervision for Della, even though she did not have capacity, what might have happened?

  Be sure to consider specific aspects of the scope of risks or harms!
- **FIND ANTECEDENT STRATEGIES**—Go back through the scenario, and draw an arrow to indicate anything you think functioned as an antecedent strategy. Make a list of the antecedent strategies, and evaluate if they were effective at preventing things like risks or actual harms. How do you know? Could anything have been improved or added?
- **IDENTIFY RED FLAGS**—Underline any red flags that you find, even if Della or Rochelle did not notice them. If they did not notice them, write down what factors may have contributed to their ethical blindness (e.g., time constraints, lack of knowledge, common practice, perceiving that there was no risk, lack of oversight). Can you think of any others?
- **FIND INSTANCES OF SOMEONE ADDRESSING THE ISSUE**—Put an * next to any language indicating that Della took specific action to address the issue. Make a list of those actions and then describe if they were effective or not, how you know, and any-

thing you would have changed or added. Were there other points in the scenario where you think Rochelle could have taken a specific action to address the situation? If so, what, what would that look like, and how would you know if it was successful?

- **CONNECT**—Have you ever been in a similar situation (as the supervisor, supervisee, or trainee) or known someone who was? What was the outcome? What went well? What would you do differently, or recommend the other person do differently, now?

Taking Care of Business—Again, your list of actions for improving your supervisory practices might still be long, and that's okay. Use the rest of this week to evaluate your list and prioritize areas that pose the biggest risk to you and your supervisees and trainees. And you might be feeling a little panic as this year spent together working through ethics content is ending. But remember, it's about incorporating ethics into your daily practice.

So, take a moment now to plan out time to work on those items that you have left. And yes, they might span out to a few months from now. And you might even consider adding this topic to your recurring agenda items with your supervisor or manager, to review your plan and ensure you're using it when you need to. You might also look at including it as a recurring agenda item with your supervisees and trainees to also review the plan in case the supervisory relationship is interrupted or needs to be transitioned or terminated. These conversations might occur quarterly or biannually.

Words of Inspiration—When someone is placed in a supervisory position, an unspoken message is communicated: "Be more like this person." So, be more like the supervisor that you want to be and believe that you are, through increased awareness, thoughtful self-reflection, and behaving with integrity. Every. Single. Day.

Weeks 50, 51, & 52—That's a Wrap

Week 50—Catching Your Breath

Week 50! You've made it through 49 weeks of content and 18 biweekly pairs of ethics topics. Wow! That's not to say that once you finish the last page of this book, you will be done thinking and talking about or reflecting and acting on ethics topics. Rather, you are in the home stretch of building fluency with the practice of keeping ethics in the forefront of your professional work. You have basically spent the last year (can you believe that?) leaning into and thoughtfully engaging with topics and thoughts and conversations and actions that many of us actively avoid.

You've worked hard to develop your ethical muscle memory; to commit to practicing something that is often challenging and uncomfortable. You've taken some breaks along the way to rest and recharge. Hopefully, in addition to building your daily practice of engaging with ethics topics, you have also built your skills related to self-compassion and self-care, to plan and manage your well-being.

For this week, all we want you to do is breathe. That's it. Just breathe. In fact, how about you take a few slow, steady, deep breaths right now? We will wait. We don't want you to actively reflect on ethics content this week. If your mind wanders to ethics topics or to-dos, just watch those thoughts pass; don't hang onto them. This week, just focus on yourself and engage in as many of your favorite self-care activities as you can cram into 7 days. We will see you next week to wrap up and provide some ideas for staying the course with your daily practice.

In case you need some ideas, we offer some of our favorite activities that keep us centered:

- ***sipping on a delicious latte***
- ***baking a loaf of bread***
- ***trying a new recipe***
- ***browsing a plant nursery (and let's be honest, coming home with another plant or two)***
- ***slowly, and mindfully, taking care of our collective 4,267 house plants***
- ***hiking***
- ***taking a ski day***
- ***taking a slow walk with the dogs and noticing small, beautiful things***
- ***taking a drive and blasting '80s music***
- ***reading or listening to poetry***

Week 51—Catching Up

You're in the home stretch now! We hope that we've been clear that the steps (and rests) you take along this journey are the important part. There's no final destination with committed practice. We've asked a lot of you along this journey. This week is just to play catch up with lingering tasks. Whether you've got tasks you can clear in a short time (say 15–30 minutes or less) or tasks that need more time and attention (i.e., you need to spend a bit of time thinking about and then get them on your calendar to work on in the coming months), this week is meant to give you time to get things in order.

We aren't going to ask you to reflect on, notice, or talk about ethics stuff. Although we not-so-secretly hope that doing those things is getting to be second nature by now. Our only ask for this week is that you don't add new items to your active to-do list or calendar. If you have ideas for future tasks or projects, put them on a list somewhere to come back to.

Here is one way you might consider using this catch-up week:

- Start this week by looking at your current to-do list and flipping back through the biweekly pairs to catch any tasks from your notes that fell through the cracks. This is a good time to get them on your list.

- Now take some time to organize your list by grouping related items and putting them in order. Some items might need to go to the top of the list because of priority or because they have to be completed before you can work on other items.
- Scan your list to see if there is anything you might be able to delegate to someone else to either get done or get started (e.g., have your supervisees and trainees do a review of their current documentation of their supervision). Make sure that you are delegating things that are within the individual's skill set.
- Now review your list and look for things that can be completed in a reasonable amount of time, say 30 minutes or less, and prioritize them for this week if you can.
- Finally, review tasks or projects that will take longer, and decide if you should get them started now or calendar them for the future. For projects or multi-step tasks, identify the first two steps, focus on getting the first step done this week, and calendar the second step. Then, before starting the second step, identify the next task and get that on your calendar. Continue in this way until you get to the last step, then celebrate!

Well, what are you waiting for?!?! Get after it, and we will see you next week.

Week 52—Reflecting & Planning

Celebration—First things first. CONGRATULATIONS!! 🎆 You made it through a ton of sometimes tricky and even unpleasant content, all in service of being the best behavior analyst you can be, and we couldn't be prouder of you. We hope that you are feeling proud of yourself. This week we want you to spend some time looking back at all you have accomplished, and then spend some time planning for the future. You'll use Monday and Tuesday to look back. Then on Wednesday, we want you to just move through your day and see how naturally you notice, think and talk about, and act on ethics topics. On Thursday and Friday, we'll prompt you to plan forward.

Monday & Tuesday—Look Back

If you can, set aside 30–60 minutes each day to review the biweekly pairs and your notes, and reflect on the content. Follow the prompts below:

- Skim over Chapters 1–3. Did you make any notes in those sections? Can you remember how you felt and what you were thinking when you read those sections? How does that compare to how you feel and what you think now?

- Flip through the biweekly pairs. Which topics were the most challenging, and why? Which were the easiest, and why?

- After flipping back through the topics, which one or two resulted in the most discomfort for you? Which ones produced the most surprising realizations about your own learning history? Why? What about the most surprising realization about your professional behavior and repertoires? Why?

- Which topic or topics produced the most surprising realization about your colleagues' behavior? Why? How about, which one resulted in the biggest change in the workplace, and why?

- Which topics resulted in the most challenging conversations with others? Why? How did those discussions turn out? And what were the outcomes of those conversations?

- Were there any conversations that you avoided? Why? What steps can you take to have them now or at least increase the chances you'll have them in the future?

- Take a few minutes and try to describe how your thoughts and behavior about ethics topics have changed over the course of developing your practice. Has your comfort level changed? Do you think and talk about ethics more frequently? Has your approach to thinking about and addressing possible ethical concerns changed over the course of your practice?

Wednesday—Notice

Spend the day just trying to notice when your thoughts, conversations, reflections, or actions center on or are related to ethics. Did you have to prompt yourself to focus on ethics, or do you feel that it came naturally, fluidly? If you think back to how often you actively engaged with ethics topics before picking up this book and how often you engage with them now, how has that changed for you? Moving through the content in the book, is there a point you can identify when you stopped having to put effort into holding ethics at the forefront of your professional activities?

Thursday & Friday—Plan Forward

As we mentioned early on in this book, we purposefully did not cover all the possible ethics topics. Why? Well, dang, that would have taken way more than 52 weeks of content. And remember, the idea was not to cover the standards in the *Ethics Code for Behavior Analysts* (BACB, 2020a).

For sure, you should develop and maintain familiarity with the standards in your professional codes of ethics. But the focus of this book was to help you deepen your relationship with ethics topics and to commit to engaging in daily practice in service of valuing ethics in our professional practice. You might be asking yourself, *Okay. Well, Emily, Sarah, and Tyra, what do I do now?* That's simple. You keep going. You continue to build your ethical muscle memory by engaging in your ethics practice. Here are some prompts and ideas to encourage you to plan forward:

- Keep working that to-do list! Don't get mad at yourself if you need to pause or push projects or tasks. Slow and steady wins the race.
- Keep an inspiration list for ethics-related ideas that come up as you continue to engage in daily ethics practice (thinking, noticing, talking, acting, reflecting). Consider storing the list in a place where your colleagues, supervisees, and trainees can add their

thoughts and ideas. Then plan regular reviews of the ideas to see if any need to move to planning and action.

- Make a list of other ethics topics that are relevant to your work context or are just plain interesting to you. Maybe they are topics that we did not cover at all. Maybe they are versions of topics that we covered. You can use the same framework we used for the biweekly pairs to engage with the topics on your list. Maybe you want to keep focusing on one topic every 2 weeks. Or maybe you want to slow down a bit and take a month to cover the topic. Either way, you are in a good position to continue the good work of building your skills with a variety of ethics topics.
- Identify any topics that you want to revisit. As you were looking back at the past 52 weeks, were there topics that you identified as uncomfortable or unfamiliar, or that you just need to spend more time with? You might intersperse these in the new topics you've identified.
- If you went through the book on your own, consider going back through it with a group of colleagues, supervisees, or trainees. You can go through start to finish, or you could pick and choose topics that seem more relevant to the needs of the group.
- Go back to Chapter 3, and flip to the end of that chapter to find the "Prepping for Identifying Professional Growth Opportunities" section. Reread that section, and think about your skills related to ethics. Are there topics that would be beneficial for you to focus on? This book has focused heavily on ethics topics that are most relevant in clinical practice.

Maybe you need to hone your ethics knowledge and skills related to billing, leadership, research, or marketing. How are your skills with taking a structured problem-solving approach or having difficult conversations? Spend some time making a list of knowledge and skills you'd like to maintain or add, and then start making your professional development plan. Are there articles you can read? Podcasts you can listen to? Workshops or presentations that you can attend? How will you allocate your time and other resources (funds, required continuing education) to executing your professional development plan?

Goodbye & Good Luck—Well, that's it. Our time together has come to an end. We hope that as you reflect on your time engaging with this book, you find that your comfort level with ethics content has increased and that you are coming away with a deeper understanding and a strong commitment to keep up with your daily ethics practice.

You are making our profession better by developing your fluency with ethics topics and by modeling for others how to take an active approach to ethics. Your journey with us probably hasn't always been easy or fun. We are grateful for the time and effort you have put into building your practice, and we hope that the content and activities serve you well throughout your career.

Much love and respect,
Tyra

With kindness and grace,
Emily

You've got this!
Sarah

Appendix A

DAILY ETHICS CROSSWALK

Scan the QR code below to access a crosswalk document that indicates the relevant standards from the current BACB ethics codes for each of the biweekly pairs in the book.

Appendix B

BOOK CLUB KIT

WELCOME

Hello there!

We are so glad that you are considering a book club for *Daily Ethics: Creating Intentional Practice for Behavior Analysts*. Or maybe you weren't really considering a book club and you were just curious about what we included in this book club kit. Cool, let us tell you why a book club can be so impactful for getting the most out of *Daily Ethics*. It's simple. Ethics, especially applied ethics, is a scary topic, and we are always at risk of delaying or avoiding altogether engaging with content that is scary or difficult. Doing challenging things with others creates safety, support, and space to share the burden of navigating the prickly parts.

Whether this is the first time you're planning on reading *Daily Ethics* and you want to go through it with a group, or you've read it and now you want to do a club to connect more deeply with the content or support others in engaging with the content, we are grateful that you are considering a book club. To help out, we have provided this kit. You'll find some considerations for organizing the book club and getting started. Then you'll find specific ideas for discussion and activities to get people engaged.

We hope that you find the book club kit helpful as you work through *Daily Ethics* or lead others through connecting with the content.

Cheers,

Tyra Emily Sarah

BOOK CLUB PREP

Step 1: Define the Purpose

The first thing to decide is the purpose of the book club. What do you want to get out of it? What do you want others to get out of it? We can think of three main purposes:

- **Personal Growth**—Are you most interested in developing your skills related to thinking and talking about ethics content? In being challenged to think about things from a different angle? Well, then your purpose is personal growth, and it is a great one! For this purpose, you can likely get away without reading the book ahead of time. In fact, contacting the content for the first time with your book club members can be very valuable, as you all will be having honest, immediate reactions to the topics and prompts.
- **Organizational Growth**—Maybe you are most interested in creating a culture where ethics is considered all the time by people. Maybe you have some concerns with the current ethical culture or practices and think that the content in the book could facilitate meaningful conversations leading to improvements. Using this book to serve this purpose may make some of the ideas and resulting tasks more acceptable to those in your workplace. You can blame the need to make changes on us! If this is your goal, then we suggest that you read through the book ahead of time so that you can be proactive in managing topics that might be triggering or difficult based on the people in your organization or recent or historical events in the workplace or in the local professional culture.
- **Supervisee/Trainee Growth**—The third common purpose of having a book club to discuss the content is to provide a growth opportunity to supervisees, trainees, or students. Conducting a book club for this purpose is so valuable, as you will be exposing individuals to this sticky content and the idea of developing a daily practice of thinking about ethics while they are in a supported context and before they have a long history of avoiding ethics, treating it as an afterthought, or wielding it as a weapon. As we suggest for groups focusing on the growth of an organization, to be best prepared to support your groups of supervis-

ees or trainees, it's important that you go through the book on your own first.

Step 2: Identify the Players

Knowing the purpose of the book club should help immediately clarify who you want to include. Here are some ideas to consider for the three main purposes:

- **Personal Growth**—If you're focusing on growing your skills related to the content, you'll likely want to invite colleagues and maybe mentors who can challenge you and facilitate expanding your perspective. We highly recommend keeping the group small (e.g., 3–5 people), including people outside of your current work setting, and having good diversity among the members.
- **Organizational Growth**—If you want to use the book as a vehicle to potentially effect change in your organization, you'll want to select individuals in decision-making roles, or at least those in positions to propose ideas to leaders. For example, if your organization has a formal ethics or training committee or workgroup, forming a book club with those individuals can be a catalyst for reviewing current processes and proposing new ideas.

  You might be tempted to include everyone, but if the committee or workgroup is large, be careful. The larger the group, the more management you'll have to do, and you run the risk of limiting meaningful participation. You know the old saying, "Too many cooks in the kitchen," right? To make things more manageable and productive, you might consider two or three smaller groups who then briefly meet every month to share, make decisions, and coordinate. You might also consider including non-behavior analysts in the book club, provided they are involved in activities with heavy ethics content (e.g., compliance, marketing, billing and documentation, human resources, operations).
- **Supervisee/Trainee Growth**—This one is obvious! You'll invite your supervisees or trainees. Unlike our suggestion for keeping the size small when focusing on organizational growth, if you structure the book club well, you can likely get away with much larger numbers in a group of supervisees or trainees. They likely have recent experience with group instructional settings (e.g.,

training, classes, group supervision), so they might not find a group setting off-putting.

Another benefit of slightly larger numbers is that you increase the number of perspectives, experiences, and exemplars that can be shared. Members can also role-play and practice giving each other feedback. Finally, assuming you create psychological safety, interesting and challenging conversations will likely occur, and you can facilitate and shape them, preparing supervisees and trainees for these conversations in the future.

But you'll have to be organized and proactively develop a culture of compassion and courage. If you create subgroups, be mindful of the makeup so that you don't group all the chatty people together or all the people who went through the same programs together. To the best of your ability, thoughtfully group people based on how they will interact in ways that bring out the best in each other.

Step 3: Decide on Logistics

Once you know your purpose and who will be invited, you need to organize the book club to make it manageable and enjoyable! Some things should probably be decided ahead of time; other things can be decided by the group, depending on the makeup. Here are a host of considerations, in no particular order:

- **Modality of Meetings**—Decide if meetings will be in person or via video chat. Video chat might allow for greater participation and easier scheduling, so those are pluses. On the other hand, people may be more likely to multitask, withdraw in the face of challenging or uncomfortable content, and feel more comfortable canceling.
- **Frequency of Meetings**—Decide how often and for how long you will meet. This will likely be informed by the purpose and makeup of the book club. For example, if you are meeting with friends and colleagues for the purpose of personal growth, you might meet less frequently but for longer. Meeting once a month or every other month might be all you can fit into your already jam-packed schedules, and meeting for an hour and a half or two might allow for some non-book-related socializing.

On the other hand, if the meetings are meant to be part of your work activities with coworkers or supervisees/trainees, you might be able to schedule them more frequently (e.g., every other week). Because the main content of the book is written in biweekly pairs, meeting every 2 weeks might be a useful cadence.

No matter the frequency of meetings, be flexible with the potential need to put things on pause anytime that is needed. Engaging with this content is challenging enough. If home or work life gets really challenging for the group, just hit pause and schedule a time to restart.

- **Reading Expectations**—You'll want to decide ahead of time, or in the first meeting, how many pages/pairs members should read between meetings. This will likely be driven in large part by the meeting schedule and what seems reasonable to get done between meetings.
- **Roles**—Depending on the purpose and makeup of the book club, you may or may not need roles. If your purpose is organizational or supervisee/trainee growth, you'll likely benefit from a few roles. At a minimum, you'll probably want a leader (to move folx through the content), a timekeeper (to keep folx on track), and a note-taker (to jot down ideas and tasks). And these roles can rotate across meetings and likely should.
- **Sharing Content & Facilitating Continued Engagement**—Decide on how documents will be shared and stored. We recommend using an easily accessible cloud-based storage system. If members will be sharing any personally identifiable information of clients or others, make sure the storage system meets any applicable requirements. You might also consider creating a way for members to chat with each other outside of the book club meetings. For example, you could set up a group chat channel in Microsoft Teams, Slack, or another platform. This can help keep the conversations or work on tasks going between meetings.
    - **Pro tip**: If you really want your book club well organized, consider using a free platform to help you manage your book club. For example, Bookclubs is an organization that provides a ton of resources for starting, running, or joining a book club. You can create an account and set up a book club

completely for free. It's pretty simple and allows to you invite people, create meetings and set reminders, create polls, and send messages to the book club members. Sure, you can pay for upgraded features if you want, but the free version works perfectly fine! Check it out here: https://bookclubs.com

- **Tone**—Spend some time thinking about the tone of the book club meetings. The purpose and members might drive this to some extent, but we encourage you to be flexible with this. For example, if you are meeting with friends, you might hold your meetings at each other's houses and have potlucks, or maybe at a café after a yoga class, or even while hiking.

  Within the workplace, we urge you to think outside the box of holding the book club in a conference or training room. Can you have the meeting somewhere more comfortable, like a break room, a nearby café, or maybe the area in a clinic used for play or gross motor activities when the clients are not there? If we want people to behave differently than they do in a more lecture-style, or formal meeting, context, well, it can be very helpful to have a change of scenery.

  Will you sit in chairs, or is it okay for folx to sit on the floor? Is it cool if people do things like eat, crochet, or doodle during the book club meetings? Can people bring homemade snacks for others? Maybe you can have the meeting in the morning while making pancakes or having an oatmeal bar? Just get creative to create a comfy space for folx to work on developing their practice with ethics topics.
- **Ground Rules**—Developing some simple ground rules that everyone can agree to at the outset is a great antecedent strategy to facilitate psychological safety. Starting the first meeting by outlining the expectations for proceeding with the book club meetings can help make people feel more comfortable. Here are some possible ground rules to consider and make your own:
  - We are here to learn and grow, not to make other people feel bad or wrong. We will value everyone's perspective and move from a place of compassionate curiosity. We will ask thoughtful questions to learn more about how people came

to their perspectives. We will respectfully share our perspectives from a desire to share, not to convince.
  - If we are doing this right, our perspectives will likely be challenged, and we will accept those challenges in service of our growth and the growth of others.
  - Our feelings may get hurt. We may hurt someone else's feelings. And we commit to discussing hurts openly and compassionately.
  - This space is for our growth, and we will not share things outside of this group.
  - We have all made mistakes. We will all make mistakes. We will frame mistakes as beautiful learning opportunities that we can share and benefit from.

Advanced Tips

For those of you who teach ethics, direct a practicum or fieldwork experience program, manage the training for your organization, or lead an ethics committee, you are likely going to repeat your book club for future courses or cohorts. To help with that, we provide two advanced tips:

1. Take a structured approach to developing and testing out the first iteration of your book club. Make an outline, use agendas, and keep running notes of what worked and what did not. You might even consider creating a survey to get feedback from each group so you can continue to make updates to the book club. From that information, develop the tools needed to efficiently run future book clubs. For example, you might want an agreement or written ground rules, pre-developed templates, scripts for introducing content or activities, or feedback forms.
2. Use a train the trainer model if you can. Consider identifying students, supervisees, or trainees to eventually lead future iterations of the book club. While they are going through the book club as a member, meet with them individually to train them to lead or co-lead meetings so that you can evaluate their performance and provide feedback. Consider developing a book club manual that leaders of book clubs can use to organize and run

the meetings. This increases the likelihood that members of the book club will have similar experiences.

Extras From the Authors

When the three of us work together, it usually involves food. So, we thought we'd share a few of our favorite dishes with you at the QR code below.

Book Activities

Discussion Questions Before Reading the Book

Use these questions to facilitate a discussion before the book club starts reading the book. Or you can send the questions out ahead of time and ask members to reflect on the questions, or maybe even answer them in a poll that you can then share back to the group.

1. What do you know about this book? Do you have any specific expectations or goals for reading it?
2. What do you know about the authors? Do you feel comfortable with them guiding you through this content?
3. What training have you had related to ethics?
4. What are your thoughts and feelings about ethics?
5. How often do you specifically think about ethics?
6. Do you think it is important to develop a daily practice related to thinking about, noticing, and taking action related to ethics in your professional activities? Why or why not?

Discussion Questions for Chapter 1: Introduction—Get the Context

Use these questions to get the members engaged right off the starting line. Again, it might be best to provide the questions ahead of time so that people can prepare if that makes them feel more comfortable participating.

1. Were you aware of how new professional ethics is in the profession of ABA-based services?

2. Would you describe yourself as more of a "black and white" thinker or a flexible thinker when it comes to ethics?
3. Do you agree with the authors that whereas professional codes of conduct are necessary to provide some rules and some guidance, it is up to individuals and the profession to interpret and apply that information to their practice? Why or why not?
4. How did the thought of developing your ethical muscle memory through creating a daily practice sit with you?
5. Why do you think the authors frame the content in the book as engaging in "practice"? Is that perspective helpful to you? Why or why not?
6. The authors make a point of saying that they are going to have readers primarily focus on self-reflection and changing the reader's own practices and behavior, as opposed to focusing on identifying instances of other people engaging in misconduct. Why do you think that is? Does that resonate with you? Are you a little uncomfortable at the thought of engaging in honest self-refection and self-evaluation?

Discussion Questions for Chapter 2: How to Use This Book—Get Your Bearings

1. Do you feel like you have a good feel for how the main part of the book will be structured?
2. What are your thoughts about spending 2 weeks on a single topic? Is there anything that concerns you?
3. The authors intentionally did not follow the outline of the sections and standards in the *Ethics Code for Behavior Analysts* (BACB, 2020a). Do you have a good sense as to why they made that decision? Can you paraphrase it? Do you agree with their approach?

Discussion Questions for Chapter 3: But First—Get Yourself Prepped

This chapter is a bit long, with a lot of prompts to check out additional resources and to engage in self-reflection. You might want to spend two meetings covering the subtopics.

1. Do you have any concerns with developing a daily ethics practice? Why or why not?

2. If you do have concerns, what are the barriers you are worried about?
3. Of the strategies suggested, which one or two are most appealing to you, and why? Which are the least appealing, and why?
4. What is your experience with engaging in structured self-reflection? Is this something you were taught or had modeled for you?
5. Is there anything that worries you about engaging in self-reflection? For example, do you think that you are a fair and accurate assessor of your own covert and overt behavior?
6. Of the strategies listed, which ones did you, or will you, try? Why?
7. If you tried some already, which ones and how did it go?
8. How do you think having compassion for others fits into our professional work? How about compassion for ourselves?
9. Does thinking or talking about compassion feel comfy or uncomfy for you? Why?
10. Did you receive any training on compassion in your professional work in grad school, supervised field work, or professional development activities through your employer?
11. What are your thoughts related to self-care?
12. Do you regularly engage in self-care activities? Do you train your supervisees and trainees to attend to their self-care needs?
13. What is your experience with professional training or education related to biases and culturally responsive practices?
14. What is your comfort level with content related to biases and culturally responsive practices?
15. Were you familiar with any of the resources shared? Did you explore any of them? Share your impressions.
16. How do you predict the topics of biases and culturally responsive practices will show up in the biweekly pairs?
17. What is your familiarity with values identification?
18. How do you think knowing your values can help you navigate ethics topics? How can they help you stay committed to engaging in daily practice?
19. Had you listed your values in the past? If so, did they change?
20. Which of the questions in the "Knowing Your Values" subsection resonated with you most, and why? Which were the most challenging to answer, and why?

21. If you are comfortable sharing, what is your current list of values?
22. Do you foresee your values changing as you work through the book?
23. What is your experience with challenging conversations? Have you been on the receiving end? How about on the initiating end? How did those conversations go?
24. What is your standard approach or response to tricky conversations? Do you run for the hills or lean in?
25. How do you see challenging situations coming up as you work through the biweekly pairs of ethics content and work on creating your daily practice?
26. Did any of the strategies provided resonate with you? Which ones and why?
27. Why do you think the authors included a section calling your attention to the need to watch for professional growth opportunities?
28. What is your typical approach to managing your professional development activities? How does that work for you?
29. Do you think that you'll identify specific strengths and areas for improvement relative to ethical practices as you move through the book? Can you predict what those areas will be?
30. What thoughts or emotions, if any, came up for you as you read this chapter?
31. Do you agree with the authors' framing of this as a journey?

Discussion Questions for Week-by-Week Content

Use some or all the questions to help facilitate deep and meaningful discussion after completing one or more of the biweekly pairs.

1. Did the biweekly topic challenge your views? Why or why not? If yes, in what ways?
2. Were you able to relate the biweekly content to your own life experiences?
3. Did the biweekly content make you feel uncomfortable about your past or current behavior? About others' past or current behavior?
4. What did you find most surprising or new in the biweekly content?

5. Was there anything in the biweekly content you disagreed with? Why?
6. Did the biweekly content motivate you to change any of your own thoughts, perspectives, or practices? If so, what, and in what ways will you make the changes?
7. Did the biweekly content motivate you to approach leaders in your organization to have discussions about things to potentially improve or add to the organization's practice or resources? If so, what, and in what ways will you make the changes?
8. Were there any "Talk About It" items that people did not get to or want to discuss?
9. Did you have to have uncomfortable conversations related to the biweekly content? If so, how did the conversations turn out?
10. How did you relate to the scenario? How did you answer the prompts?
11. What actions did you take during the biweekly topic, and what tasks ended up on your ongoing to-do list?

Additional Activities

Use these activities to further enhance members' connection to the content and to develop useful resources.

- Offer members the opportunity to share an example of a past mistake or issue related to the biweekly content. It can be a mistake or misstep they made, or one they saw someone else make. Talk through how a more nuanced understanding of ethics and engaging in daily ethics practice could have resulted in a different outcome.
- Develop scripts for addressing red flags or concerns about misconduct or a possible ethics violation related to the biweekly content, and role-play having the difficult conversations.
- Have a rotating role to read the biweekly scenario and ask members to share how they answered the prompts.
- Develop some scripts together and role-play parts of the scenario.
- Divide and conquer by sharing to-dos, and then create subgroups to work on completing tasks and sharing back to the group.

Discussion Questions After Reading the Whole Book

1. What did you know about this book before getting started?
2. Why did you want to read this book? What were your expectations? What were you hoping to get out of it? Did you get the outcome you were looking for?
3. What surprised you the most about the content in the book? What surprised you the most about how you reacted to the content in the book?
4. In what ways did your thoughts and actions change after reading the book?
5. What were some of your private events while reading the book? Did the content evoke any emotions? Make you laugh, cry, or cringe? Did any of the content make you feel proud or guilty?
6. Were you successful at developing a daily ethics practice? At what point in the book did your daily practice become easier and more fluent?
7. What strategies were most helpful in cultivating your daily ethics practices?
8. What are your two top favorite biweekly pairs?
9. Which biweekly pair was the most challenging for you, and why? How about the easiest, and why?
10. What are some of the biggest changes you made to your practice and/or your organization's practice during or after reading the book?
11. Did you highlight or bookmark any passages from the book? Did you have a favorite quote or quotes? If so, share which ones and why.
12. Would you recommend this book to another behavior analyst? Why or why not?

Discussion Questions About the Authors

1. Why do you think the three authors wrote this book? What was their purpose? Can you describe an overarching theme or themes?
2. If you could sum up the book in a few words or sentences, what would they be?

3. Do you think the authors' personal views or biases influenced the way the content was presented? If so, in what ways? Where was it most evident?
4. How does this book compare to other books on ethics in behavior analysis?
5. If you could ask the authors one question about this book, what would it be?
6. If you could ask the authors to change, take out, or add one thing, what would it be?

References

Afolabi, O. E. (2015). Dual relationships and boundary crossing: A critical issues in clinical psychology practice. *International Journal of Psychology and Counselling, 7*(2), 29–39.

Allen, S. (2018a). *The science of awe* [White paper]. Greater Good Science Center at UC Berkeley. https://ggsc.berkeley.edu/images/uploads/GGSC-JTF_White_Paper-Awe_FINAL.pdf

Allen, S. (2018b). *The science of gratitude* [White paper]. Greater Good Science Center at UC Berkeley. https://ggsc.berkeley.edu/images/uploads/GGSC-JTF_White_Paper-Gratitude-FINAL.pdf

American Speech-Language-Hearing Association (n.d.). *Cultural responsiveness.* Retrieved November 21, 2023, from https://www.asha.org/practice-portal/professional-issues/cultural-responsiveness/

Association of Professional Behavior Analysts. (n.d.). *Member resources.* https://www.apbahome.net/page/memberresources

Auteri, S. (2014, February). *Should dual relationships always be off-limits?* AASECT. https://www.aasect.org/should-dual-relationships-always-be-limits

Bailey, J. S., & Burch, M. R. (2019). *Analyzing ethics questions from behavior analysts: A student workbook.* Routledge. https://doi.org/10.4324/9781351117784

Bailey, J. S., & Burch, M. R. (2020). *The RBT® Ethics Code: Mastering the BACB© ethical requirements for Registered Behavior Technicians™*. Routledge. https://doi.org/10.4324/9780367814922

Bailey, J. S., & Burch, M. R. (2022). *Ethics for behavior analysts* (4th ed.). Routledge. https://doi.org/10.4324/9781003198550

Bazerman, M. H., & Gino, F. (2012). Behavioral ethics: Toward a deeper understanding of moral judgment and dishonesty. *Annual Review of Law and Social Science,* 8, 85–104. https://doi.org/10.1146/annurev-lawsocsci-102811-173815

Beaulieu, L., & Jimenez-Gomez, C. (2022), Cultural responsiveness in applied behavior analysis: Self-assessment. *Journal of Applied Behavior Analysis,* 55(2), 337–356. https://doi.org/10.1002/jaba.907

Behavior Analyst Certification Board. (n.d.). *BACB certificant data.* Retrieved December 12, 2023, from https://www.bacb.com/bacb-certificant-data/

Behavior Analyst Certification Board. (2001). *Guidelines for responsible conduct for behavior analysts.* Retrieved March 2, 2024, from https://www.bacb.com/wp-content/uploads/2020/09/2001-Conduct-Guidelines.pdf

Behavior Analyst Certification Board. (2018a). *Consulting supervisor requirements for new BCBAs supervising fieldwork.* https://www.bacb.com/bcba/

Behavior Analyst Certification Board. (2018b). *A summary of ethics violations and code-enforcement activities: 2016–2017* [White paper]. Author. Retrieved November 18, 2023, from https://www.bacb.com/ethics-information/ethics-resources/

Behavior Analyst Certification Board. (2019). *Fieldwork checklist and tip sheet.* Retrieved November 14, 2023, from https://www.bacb.com/bcba/

Behavior Analyst Certification Board. (2020a). *Ethics code for behavior analysts.* Retrieved November 14, 2023, from https://www.bacb.com/ethics-information/ethics-codes/

Behavior Analyst Certification Board. (2020b). Taking supervision seriously. *BACB Newsletter.* Retrieved December 30, 2023, from https://www.bacb.com/wp-content/uploads/2020/08/BACB_August2020_Newsletter-230718-a.pdf#page=5

Behavior Analyst Certification Board. (2021a, August 10). *2022 supervised fieldwork summary* [Video]. YouTube. Retrieved December 30, 2023, from https://www.bacb.com/documenting-fieldwork-hours/

Behavior Analyst Certification Board (2021b). *Continuity of services: Managing service interruptions, transitions, and discontinuations.* Retrieved December 30, 2023, from https://www.bacb.com/ethics-information/ethics-resources/

Behavior Analyst Certification Board (2021c). *Continuity of services: Reminders for RBTs.* Retrieved November 14, 2023, from https://www.bacb.com/ethics-information/ethics-resources/

Behavior Analyst Certification Board. (2021d). *RBT ethics code* (2.0). Retrieved November 14, 2023, from https://www.bacb.com/ethics-information/ethics-codes/

Behavior Analyst Certification Board. (2022a). *BCaBA handbook.* Retrieved November 14, 2023, from https://www.bacb.com/bcaba/

Behavior Analyst Certification Board. (2022b). *BCBA handbook.* Retrieved November 14, 2023, from https://www.bacb.com/bcba/

Behavior Analyst Certification Board. (2022c). *RBT handbook.* Retrieved November 14, 2023, from https://www.bacb.com/rbt/

Behavior Analyst Certification Board. (2022d). *Supervision checklist for RBTs.* Retrieved December 30, 2023, from https://www.bacb.com/rbt/?rbt-supervision#rbtResourceCarousel

Behavior Analyst Certification Board. (2022e). *Supervision checklist for RBT supervisors and RBT requirements coordinators.* Retrieved December 30, 2023, from https://www.bacb.com/rbt/?rbt-supervision#rbtResourceCarousel

Behavior Analyst Certification Board. (2023a, October 12). *Maintaining your RBT certification* [Video]. https://www.bacb.com/maintaining-your-rbt-certification/

Behavior Analyst Certification Board. (2023b). *A summary of ethics violations and code-enforcement activities: 2019-2021* [White paper]. Author. Retrieved November 14, 2023, from https://www.bacb.com/ethics-information/ethics-resources/

Behavioral Healthcare Providers. (2016). *Life changes stress test.* https://www.bhpcare.com/patient-information/self-management-tools/life-changes-stress-test/

Beirne, A., & Sadavoy, J. A. (2021). *Understanding ethics in applied behavior analysis: Practical applications* (2nd ed.). Routledge. https://doi.org/10.4324/9781003190707

Breaux, C. A., & Smith, K. (2023). Assent in applied behaviour analysis and positive behaviour support: Ethical considerations and practical recommendations. *International Journal of Developmental Disabilities*, *69*(1), 111–121. https://doi.org/10.1080/20473869.2022.2144969

Brodhead, M. T., & Higbee, T. S. (2012). Teaching and maintaining ethical behavior in a professional organization. *Behavior Analysis in Practice*, *5*(2), 82–88. https://doi.org/10.1007/BF03391827

Brodhead, M. T., Quigley, S. P., & Cox, D. J. (2018). How to identify ethical practices in organizations prior to employment. *Behavior Analysis in Practice, 11*(2), 165–173. https://doi.org/10.1007/s40617-018-0235-y

Brodhead, M. T., Quigley, S. P., & Wilczynski, S. M. (2018). A call for discussion about scope of competence in behavior analysis. *Behavior Analysis in Practice, 11*(4), 424–435. https://doi.org/10.1007/s40617-018-00303-8

Brodhead, M. T., Cox, D. J., & Quigley, S. P. (2022). *Practical ethics for effective treatment of autism spectrum disorder* (2nd ed.). Academic Press. https://doi.org/10.1016/C2020-0-00653-5

Brownstein, M. (2019). *Implicit bias.* Stanford Encyclopedia of Philosophy. https://plato.stanford.edu/entries/implicit-bias/

Butler, L. D. (n.d.). *Developing your self-care plan.* University at Buffalo School of Social Work. https://socialwork.buffalo.edu/resources/self-care-starter-kit/developing-your-self-care-plan.html

Campbellsville University. (2020, October 8). *Navigating dual relationships in social work.* https://online.campbellsville.edu/social-work/dual-relationships-in-social-work/

Carnegie, D. (2022). *How to win friends and influence people.* Gallery Books.

Čolić, M., Araiba, S., Lovelace, T. S., & Dababnah, S. (2022). Black caregivers' perspectives on racism in ASD services: Toward culturally responsive ABA practice. *Behavior Analysis in Practice, 15*(4), 1032–1041. https://doi.org/10.1007/s40617-021-00577-5

Connors, B. M., & Capell, S. T. (Eds.). (2020). *Multiculturalism and diversity in applied behavior analysis: Bridging theory and application.* Routledge.

Cox, D. J. (2020). A guide to establishing ethics committees in behavioral health settings. *Behavior Analysis in Practice, 13*(4), 939–949. https://doi.org/10.1007/s40617-020-00455-6

Cozma, I. (2023, February 07). How to find, define, and use your values. *Harvard Business Review*. https://hbr.org/2023/02/how-to-find-define-and-use-your-values

Donner, M. B., VandeCreek, L., Gonsiorek, J. C., & Fisher, C. B. (2008). Balancing confidentiality: Protecting privacy and protecting the public. *Professional Psychology: Research and Practice, 39*(3), 369–376. https://doi.org/10.1037/0735-7028.39.3.369

Drumwright, M., Prentice, R., & Biasucci, C. (2015). Behavioral ethics and teaching ethical decision making. *Decision Sciences Journal of Innovative Education, 13*(3), 431–458. https://doi.org/10.1111/dsji.12071

Fiebig, J. H., Gould, E. R., Ming, S., & Watson, R. A. (2020). An invitation to act on the value of self-care: Being a whole person in all that you do. *Behavior Analysis in Practice, 13*(3), 559–567. https://doi.org/10.1007/s40617-020-00442-x

FitzGerald, C., & Hurst, S. (2017). Implicit bias in healthcare professionals: A systematic review. *BMC Medical Ethics, 18*(1), 19. https://doi.org/10.1186/s12910-017-0179-8

Flowers, J. (2023). *Behavior technician ethics workbook*. Beluga Publications.

Flowers, J., & Dawes, J. (2023). Dignity and respect: Why therapeutic assent matters. *Behavior Analysis in Practice, 16*(4), 913–920. https://doi.org/10.1007/s40617-023-00772-6

Fong, E. H., Ficklin, S., & Lee, H. Y. (2017). Increasing cultural understanding and diversity in applied behavior analysis. *Behavior Analysis: Research and Practice, 17*(2), 103–113. https://doi.org/10.1037/bar0000076

Foxx, R. M. (1996). Translating the covenant: The behavior analyst as ambassador and translator. *The Behavior Analyst, 19*(2), 147–161. https://doi.org/10.1007/BF03393162

Fraidlin, A., McElroy, A., Moses, K., Jenssen, K., & Van Stratton, J. E. (2023). Designing a successful supervision journey: Recommendations and resources for new BCBA supervisors. *Behavior Analysis in Practice, 16*(2), 374–387. https://doi.org/10.1007/s40617-022-00728-2

Garza, K. L., McGee, H. M., Schenk, Y. A., & Wiskirchen, R. R. (2018). Some tools for carrying out a proposed process for supervising experience hours for aspiring Board Certified Behavior Analysts®. *Behavior Analysis in Practice, 11*(1), 62–70. https://doi.org/10.1007/s40617-017-0186-8

Gatzunis, K. S., Edwards, K. Y., Rodriguez Diaz, A., Conners, B. M., & Weiss, M. J. (2022). Cultural responsiveness framework in BCBA® supervision. *Behavior Analysis in Practice, 15*(4), 1373–1382. https://doi.org/10.1007/s40617-022-00688-7

Gentile, M. C. (2012). *Giving voice to values: How to speak your mind when you know what's right*. Yale University Press.

Gibson, C., Medeiros, K. E., Giorgini, V., Mecca, J. T., Devenport, L. D., Connelly, S., & Mumford, M. D. (2014). A qualitative analysis of power differentials in ethical situations in academia. *Ethics & Behavior, 24*(4), 311–325. https://doi.org/10.1080/10508422.2013.858605

Gilbert, P. (2009). *The compassionate mind: A new approach to life's challenges.* Constable.

Greater Good. (n.d.). *Gratitude quiz.* https://greatergood.berkeley.edu/quizzes/take_quiz/gratitude

Greater Good Science Center. (n.d.-a). *Discover new practices.* Greater Good in Action. https://ggia.berkeley.edu/?_ga=2.69329168.1465280276.1700418793-1505963900.1700418793#filters=

Greater Good Science Center. (n.d.-b). Thnx4! https://www.thnx4.org

Green, K., Lewon, M., Lewon, A. B., & Ghezzi, P. M. (2023). All things must pass: Termination of services in behavior analysis [Advance online publication]. *Behavior Analysis in Practice.* https://doi.org/10.1007/s40617-023-00854-5

Grenny, J., Patterson, K., McMillan, R., Switzler, A., & Gregory, E. (Eds.). (2022). *Crucial conversations: Tools for talking when stakes are high* (3rd ed.). McGraw Hill.

Herlihy, B., & Corey, G. (1992). *Dual relationships in counseling.* American Association for Counseling and Development.

Homewood Health (n.d.). *Building a self-care plan.* https://brocku.ca/human-rights/wp-content/uploads/sites/208/unct-ye-dss-doc-building-self-care-toolkit-en.pdf

Jenkins, S., & Ulrich, R. (Hosts). (2023, October 12). *Better understanding supervision as an RBT* [Audio podcast episode]. *In Inside the BACB.* Behavior Analyst Certification Board. https://www.bacb.com/bacb-podcasts/

Jimenez-Gomez, C., & Beaulieu, L. (2022). Cultural responsiveness in applied behavior analysis: Research and practice. *Journal of Applied Behavior Analysis, 55*(3), 650–673. https://doi.org/10.1002/jaba.920

Jin, G., & Merritt, D. L. (1993). Delegating as a component of managing effectively. *The Journal of Technology Studies, 19*(2), 60–63. https://www.jstor.org/stable/43603685

The Joint Commission. (2022, April). Informed consent: More than getting a signature. *Quick Safety, 21,* 1–3. https://www.jointcommission.org/resources/news-and-multimedia/newsletters/newsletters/quick-safety/quick-safety--issue-21-informed--consent-more-than-getting-a-signature/informed-consent-more-than-getting-a-signature/

Kazemi, E., Rice, B., & Adzhyan, P. (2019). *Fieldwork and supervision for behavior analysts: A handbook.* Springer Publishing Company.

Kelly, E. M., Greeny, K., Rosenberg, N. E., & Schwartz, I. S. (2023). How behavior analysts make ethical decisions: A qualitative study [Advance online publication]. *Behavior Analysis in Practice.* https://doi.org/10.1007/s40617-023-00804-1

Kelly, A. N., Shraga, E., & Bollinger, L. (2023). *Back to basics: Ethics for behavior analysts.* Academic Press. https://doi.org/10.1016/C2020-0-02220-6

Keltner, D., & Haidt, J. (2003). Approaching awe, a moral, spiritual, and aesthetic emotion. *Cognition and Emotion, 17*(2), 297–314. https://doi.org/10.1080/02699930302297

Kirby, M. S., Spencer, T. D., & Spiker, S. T. (2022). Humble behaviorism redux. *Behavior and Social Issues, 31*(1), 133–158. https://doi.org/10.1007/s42822-022-00092-4

LeBlanc, L. A., & Karsten, A. (2024). *Ethics: Proactive and practical decision making for behavior analysts.* Sloan Publishing.

LeBlanc, L. A., Onofrio, O. M., Valentino, A. L., & Sleeper, J. D. (2020). Promoting ethical discussions and decision making in a human service agency. *Behavior Analysis in Practice, 13*(4), 905–913. https://doi.org/10.1007/s40617-020-00454-7

LeBlanc, L. A., & Sellers, T. P. (2022). *The consulting supervisor's workbook: Supporting new supervisors.* KeyPress Publishing.

LeBlanc, L. A., Sellers, T. P., & Ala'i, S. (2020). *Building and sustaining meaningful and effective relationships as a supervisor and mentor.* Sloan Publishing.

LeBlanc, L. A., Taylor, B. A., & Marchese, N. V. (2020). The training experiences of behavior analysts: Compassionate care and therapeutic relationships with caregivers. *Behavior Analysis in Practice, 13*(2), 387–393. https://doi.org/10.1007/s40617-019-00368-z

Leland, W., & Stockwell, A. (2019). A self-assessment tool for cultivating affirming practices with transgender and gender-nonconforming (TGNC) clients, supervisees, students, and colleagues. *Behavior Analysis in Practice, 12*(4), 816–825. https://doi.org/10.1007/s40617-019-00375-0

Longstaff, S. (2017). *Everyday ethics.* Ventura Press.

Luna, O., & Rapp, J. T. (2019). Using a checklist to increase objective session note writing: Preliminary results. *Behavior Analysis in Practice, 12*(3), 622–626. https://doi.org/10.1007/s40617-018-00315-4

Mathur, S. K., & Rodriguez, K. A. (2022). Cultural responsiveness curriculum for behavior analysts: A meaningful step toward social justice. *Behavior Analysis in Practice, 15*(4), 1023–1031. https://doi.org/10.1007/s40617-021-00579-3

Mead Jasperse, S. C., Kelly, M. P., Ward, S. N., Fernand, J. K., Joslyn, P. R., & van Dijk, W. (2023). Consent and assent practices in behavior analytic research [Advance online publication]. *Behavior Analysis in Practice.* https://doi.org/10.1007/s40617-023-00838-5

Merriam-Webster. (n.d.-a). Delegate. In *Merriam-Webster.com dictionary.* Retrieved June 11, 2023, from https://www.merriam-webster.com/dictionary/delegate

Merriam-Webster. (n.d.-b). Practice. In *Merriam-Webster.com dictionary*. Retrieved August 26, 2023, from https://www.merriam-webster.com/dictionary/practice

Morris, C., Detrick, J. J., & Peterson, S. M. (2021). Participant assent in behavior analytic research: Considerations for participants with autism and developmental disabilities. *Journal of Applied Behavior Analysis, 54*(4), 1300–1316. https://doi.org/10.1002/jaba.859

Neff, K. (n.d.). *Self-compassion*. https://self-compassion.org/

Neff, K. D., & Knox, M. C. (2020). Self-compassion. In V. Zeigler-Hill & T. K. Shackelford (Eds.), *Encyclopedia of personality and individual differences* (pp. 4663–4670). Springer, Cham. https://doi.org/10.1007/978-3-319-24612-3_1159

Palazzo, G., Krings, F., & Hoffrage, U. (2012). Ethical blindness. *Journal of Business Ethics, 109*(3), 323–338. https://doi.org/10.1007/s10551-011-1130-4

Parsons, M. B., Rollyson, J. H., & Reid, D. H. (2012). Evidence-based staff training: A guide for practitioners. *Behavior Analysis in Practice, 5*(2), 2–11. https://doi.org/10.1007/BF03391819

Penney, A. M., Bateman, K. J., Veverka, Y., Luna, A., & Schwartz, I. S. (2023). Compassion: The eighth dimension of applied behavior analysis [Advance online publication]. *Behavior Analysis in Practice*. https://doi.org/10.1007/s40617-023-00888-9

Personal Values (n.d.). Personal values assessment. https://personalvalu.es

Pope, K. S. (1988). How clients are harmed by sexual contact with mental health professionals: The syndrome and its prevalence. *Journal of Counseling & Development, 67*(4), 222–226. https://doi.org/10.1002/j.1556-6676.1988.tb02587.x

Pope, K. S., & Vasquez, M. J. T. (2016). *Ethics in psychotherapy and counseling: A practical guide* (5th ed.). John Wiley & Sons.

Pritchett, M., Ala'i-Rosales, S., Cruz, A. R., & Cihon, T. M. (2022). Social justice is the spirit and aim of an applied science of human behavior: Moving from colonial to participatory research practices. *Behavior Analysis in Practice, 15*(4), 1074–1092. https://doi.org/10.1007/s40617-021-00591-7

Project Implicit. (n.d.). Project implicit. https://www.projectimplicit.net/

Project Implicit. (2011). Implicit association test (IAT). https://implicit.harvard.edu/implicit/takeatest.html

Rodriguez, K. A., Tarbox, J., & Tarbox, C. (2023). Compassion in autism services: A preliminary framework for applied behavior analysis. *Behavior Analysis in Practice, 16*(4), 1034–1046. https://doi.org/10.1007/s40617-023-00816-x

Rohrer, J. L., Marshall, K. B., Suzio, C., & Weiss, M. J. (2021). Soft skills: The case for compassionate approaches or how behavior analysis keeps finding its heart. *Behavior Analysis in Practice, 14*(4), 1135–1143. https://doi.org/10.1007/s40617-021-00563-x

Rosenberg, M. B. (2015). *Nonviolent communication: A language of life* (3rd ed.). PuddleDancer Press.

Rosenberg, N. E., & Schwartz, I. S. (2019). Guidance or compliance: What makes an ethical behavior analyst? *Behavior Analysis in Practice, 12*(2), 473–482. https://doi.org/10.1007/s40617-018-00287-5

Scott, K. (2019). *Radical candor: Be a kick-ass boss without losing your humanity* (Rev. ed.). St. Martin's Press.

Sellers, T. P., Alai-Rosales, S., & MacDonald, R. P. F. (2016). Taking full responsibility: The ethics of supervision in behavior analytic practice. *Behavior Analysis in Practice, 9*(4), 299–308. https://doi.org/10.1007/s40617-016-0144-x

Sellers, T. P., & LeBlanc, L. A. (2022). *The new supervisor's workbook: Success in the first year of supervision*. KeyPress Publishing.

Sellers, T. P., LeBlanc, L. A., & Valentino, A. L. (2016). Recommendations for detecting and addressing barriers to successful supervision. *Behavior Analysis in Practice, 9*(4), 309–319. https://doi.org/10.1007/s40617-016-0142-z

Sellers, T. P., Seniuk, H. A., Lichtenberger, S. N., & Carr, J. E. (2023). The history of the Behavior Analyst Certification Board's ethics codes [Advance online publication]. *Behavior Analysis in Practice*. https://doi.org/10.1007/s40617-023-00803-2

Sellers, T. P., Valentino, A. L., & LeBlanc, L. A. (2016). Recommended practices for individual supervision of aspiring behavior analysts. *Behavior Analysis in Practice, 9*(4), 274–286. https://doi.org/10.1007/s40617-016-0110-7

Sidman, M. (1993). Reflections on behavior analysis and coercion. *Behavior and Social Issues, 3*(1 & 2), 75–85.

Sostrin, J. (2017, October 10). To be a great leader, you have to learn how to delegate well. *Harvard Business Review*. https://hbr.org/2017/10/to-be-a-great-leader-you-have-to-learn-how-to-delegate-well

Spencer, T. D., Slim, L., Cardon, T., & Morgan, L. (n.d.). *Interprofessional collaborative practice between behavior analysts and speech-language pathologists*. Association for Behavior Analysis International. https://www.abainternational.org/constituents/practitioners/interprofessional-collaborative-practice.aspx

Stone, D., Patton, B., & Heen, S. (2023). *Difficult conversations: How to discuss what matters most* (3rd ed.). Penguin Books.

Strauss, C., Taylor, B. L., Gu, J., Kuyken, W., Baer, R., Jones, F., & Cavanagh, K. (2016). What is compassion and how can we measure it? A review of definitions and measures. *Clinical Psychology Review, 47*, 15–27. https://doi.org/10.1016/j.cpr.2016.05.004

Sush, D. J., & Najdowski, A. C. (2022). *A workbook of ethical case scenarios in applied behavior analysis* (2nd ed.). Academic Press. https://doi.org/10.1016/C2021-0-00457-0

Tavoian, D., & Craighead, D. H. (2023). Deep breathing exercise at work: Potential applications and impact. *Frontiers in Physiology, 14*. https://doi.org/10.3389/fphys.2023.1040091

Taylor, B. A., LeBlanc, L. A., & Nosik, M. R. (2019). Compassionate care in behavior analytic treatment: Can outcomes be enhanced by attending to relationships with caregivers? *Behavior Analysis in Practice*, *12*(3), 654–666. https://doi.org/10.1007/s40617-018-00289-3

Tervalon, M., & Murray-García, J. (1998). Cultural humility versus cultural competence: A critical distinction in defining physician training outcomes in multicultural education. *Journal of Health Care for the Poor and Underserved*, *9*(2), 117–125. https://doi.org/10.1353/hpu.2010.0233

Thoreau, H. D. (2016). *Walden*. Pan Macmillan.

Turner, L. B., Fischer, A. J., & Luiselli, J. K. (2016). Towards a competency-based, ethical, and socially valid approach to the supervision of applied behavior analytic trainees. *Behavior Analysis in Practice*, *9*(4), 287–298. https://doi.org/10.1007/s40617-016-0121-4

Warley (n.d.). *My personal self-assessment*. Life Skills That Matter. https://lifeskillsthatmatter.com/personal-self-assessment/

Weiss, M. J., & Russo, S. (2022). Confidentiality in the age of social media. In A. Beirne & J. A. Sadavoy (Eds.), *Understanding ethics in applied behavior analysis: Practical applications* (2nd ed., pp. 349–353). Routledge. https://doi.org/10.4324/9781003190707-33

Widhalm, C., & Vernoy, K. (Hosts). (2022, November 28). It's the lack of thought that counts: Ethical decision making in dual relationships (No. 288) [Audio podcast episode]. *In Modern Therapist's Survival Guide*. Therapy Reimagined. https://therapyreimagined.com/modern-therapist-podcast/its-the-lack-of-thought-that-counts-ethical-decision-making-in-dual-relationships/

Wolf, M. M. (1978). Social validity: The case for subjective measurement or how applied behavior analysis is finding its heart. *Journal of Applied Behavior Analysis*, *11*(2), 203–214. https://doi.org/10.1901/jaba.1978.11-203

Wright, P. I. (2019). Cultural humility in the practice of applied behavior analysis. *Behavior Analysis in Practice*, *12*(4), 805–809. https://doi.org/10.1007/s40617-019-00343-8

About the Authors

Tyra P. Sellers, JD, PhD, BCBA-D, is the CEO of the Association of Professional Behavior Analysts and former Director of Ethics for the Behavior Analyst Certification Board (BACB). She earned a BA in Philosophy and an MA in Special Education from San Francisco State University, a JD from the University of San Francisco, and a PhD from Utah State University. She is a Board Certified Behavior Analyst. She co-authored the book *Building and Sustaining Meaningful and Effective Relationships as a Supervisor and Mentor* and the workbook pair titled *The New Supervisor's Workbook* and *The Consulting Supervisor's Workbook*. She has lived in many places but calls home wherever she finds herself with her partner and children.

Emily A. Patrizi, MS, BCBA, earned a BS in Elementary and Special Education from East Stroudsburg University and an MS in Applied Behavior Analysis and Autism from Sage Colleges. She is a Board Certified Behavior Analyst. She held a teaching degree within an autism support program before serving across multiple clinical and operational roles within the field. She recently held the position of Chief Operating Officer at Trumpet Behavioral Health. She has served as a BACB Ethics Subject Matter Expert and established a behavioral program within an interdisciplinary treatment model. Her most important and valued role is being a mom.

Sarah Lichtenberger, PhD, BCBA-D, earned a BA in English and Psychology from the University of North Carolina at Chapel Hill and an MA and PhD in Behavior Analysis from Western Michigan University. She is a Board Certified Behavior Analyst. She completed a postdoctoral fellowship at the Kennedy Krieger Institute in the Neurobehavioral Unit Outpatient Clinic and has held positions as a senior clinician for Trumpet Behavioral Health, the Ethics Education Manager and Assistant Director of Ethics for the BACB, and the Director of Clinical Standards for Verbal Beginnings. She is a Senior Account Manager for BehaviorLive and she serves as an Ethics Subject Matter Expert for the Association of Professional Behavior Analysts.

Explore more books from KeyPress Publishing to support your clinical practice.

Now What?
A Behavior Analyst's
First-Year Survival Guide

The Consulting Supervisor and New Supervisor Workbooks

Check here for more *Daily Ethics* resources from the authors.
dailyethicsbook.com

www.KeyPressPublishing.com

Made in the USA
Columbia, SC
20 August 2024